Gary,

Enjoy the read!

Phil

Rik Coeckelbergs

A New Ethos in Banking

Embracing Values and Ethics for a Meaningful Transformation

palgrave
macmillan

Rik Coeckelbergs
The Banking Scene
Mechelen, Belgium

ISBN 978-3-031-78233-6 ISBN 978-3-031-78234-3 (eBook)
https://doi.org/10.1007/978-3-031-78234-3

Cover illustration: eStudio Calamar

This Palgrave Macmillan imprint is published by the registered company Springer Nature Switzerland AG
The registered company address is: Gewerbestrasse 11, 6330 Cham, Switzerland

Acknowledgements

Writing a first book is a daunting challenge, and the hardest part is often saying “yes, I will”, at least that was the case for me. Over the years, many asked when I would write a book, but the idea always gave me cold feet. What could I say that hadn’t been said before? How would my work stand out? What would the world think of it? What should I even write about, and why?

In hindsight, those were just excuses to delay starting. What ultimately made me take the leap was the encouragement of Theo Lau. “Good words and good thoughts need to be shared,” she told me. Her belief in my message pushed me to take that first step, guiding me through the process of drafting a proposal and reaching out to publishers. Thank you, Theo, for that invaluable push!

Of course, I couldn’t have done it without additional support from my wife, Lies, and friends like Andrew Vorster and Leda Glyptis. Their gentle but persistent nudges kept me moving forward. I am also deeply grateful to Tula Weiss and the publishing team for believing in the project and helping me get out of my comfort zone.

Finally, my deepest thanks to Lies once again, for your endless patience and understanding during the process to crafting the book. And to my parents-in-law, where we lived for ten months during the writing process—thank you for your support in what was undoubtedly an intense period. The time I spent on this book meant less time with you and the kids, and I am deeply grateful for your encouragement and belief in me throughout this journey.

Then came the writing itself—alongside balancing a renovation, raising two (in the meantime three) small children, and running a business. It was a whirlwind, but I am incredibly thankful to everyone who supported, inspired, and encouraged me along the way. To those who reminded me why reflecting on the future of banking matters, I couldn't have done it without you.

A special thanks goes to Tim De Smedt for his dedicated review and invaluable feedback. Your countless comments sharpened the manuscript in ways that I could never have achieved alone. Tim, this book wouldn't be what it is without your input!

The theme of the book draws on the many discussed from our conferences in 2023, including, but not exclusively, inspirational talks from the following speakers: Paolo Sironi, Jerry Grbic, Isabelle Waty, Nasir Zubairi, Olga Bogdanova, Analia Clouet, Lorand Gabos Szoverdi, Helène Erftemeijer, Bas Rapati, Joris Krijger, Nathalie Hazenberg, Wies Wagenaar, Rik Goslinga, Léon Wijnands, Niccolo Polli, David Suetens, Kurt Vanhee, Thomas Van Rompuy, Ludivine Pilate, Stefaan Roegiers, Gabriela Giannattasio, Karin Van Hoecke, Nathalie Delaere, Klaas Ariaans, Wim Mijs, Guy Schellinck, Roland van der Vorst, Leen Van den Neste. We had many more enlightening conversations at other events, and though I can't list them all, they were no less valuable.

I am also deeply thankful to those who shared their stories, whether on or off the record. The interviews I conducted, along with the insights shared on our website, served as rich sources for this book. Beyond those mentioned above, my sincere gratitude goes to Michael Anseeuw, Alexandros Efstratoglou, Corin Millais, Martin Rohner, Eelco Dubbeling, Hedwige Nuyens, and Peter Surek.

To all of you who contributed, whether in big ways or small—thank you for helping shape this book into a story of the industry, not just my own narrative.

Contents

About the Author

Rik Coeckelbergs is the Founder and Managing Director of The Banking Scene, a network organisation for bankers and is passionate about banking and conceptualising how to turn the industry into a better self. Over the years, the community grew to 5 inspirational events in Luxembourg, Amsterdam and Brussels, the biggest with +400 attendees.

Before his entrepreneurial journey, Rik worked for +10 years in the financial services industry in various roles, both as an external consultant and in internal roles in business, strategy and marketing. In 2009, he created a LinkedIn community called "Innovation In Payments", which today is LinkedIn's largest payments group worldwide with +125k members.

Rik Coeckelbergs is a regular advisor, speaker and blogger on thebankingscene.com and a guest writer for third-party websites. Occasionally, Rik contributes to publications, like in 2019 in "Regtech Blackbook—The Sequel" and in 2021 in Chris Skinner's most recent bestselling book "Digital for Good".

1

Introduction: We Need to Rethink Transformation

Thank you for opening and reading this book. I feel honoured. Truth be told, it is my first book, I already felt honoured the moment a publisher asked me to write a book. That you now take the first step to read it makes me even more proud. Thank you again for giving me the opportunity to share it with you.

I have been working in the banking industry for years, constantly learning and expanding my knowledge in the field. It is an endless journey, and I genuinely enjoy it. I hope this book will also inspire you in some way. Chances are, you are also working in the banking industry. We may share a similar passion for your job and the organisation you work for. However, like myself, you may occasionally question what you are doing. You may question the tasks you are assigned to or the level of fulfilment you gain from them. You may even question the values of your organisation or the priorities that have been set to improve the bank's future.

I feel you; I've been there. I've enjoyed working in a bank. A few years back, I set up my own business, a community for the banking industry, with a mission "To connect organisations and people, helping to shape the bank of tomorrow".

I am energised by the endless conversations I have with bankers from all levels. I love to hear about their struggles and successes, and I enjoy discussing their convictions about how they can leave a bigger impact on their clients, society, and shareholders.

R. Coeckelbergs, *A New Ethos in Banking*,
https://doi.org/10.1007/978-3-031-78234-3_1

I've worked in the industry, and today, I overlook the industry from the outside by organising conferences and meetups for it. By nature, I don't think I am the person who loves all the management that comes with the organisation of events. I love the orchestration of the conversations on stage and, perhaps even more, the private conversation to prepare these events and the outcomes afterwards. The theme of the events often stimulates the speakers and attendees to take a moment to reflect.

I see that as my contribution to better banking tomorrow.

Reading this book probably means that you received a good education and probably have a decent job and income. You entered a circle with an interest in banking. You read this book partly because of the chances in life that were given to you. You grew up in a world of opportunities, the consumer economy. I believe that banks play a fundamental role in the development of the world we know today. If money makes the world go round, banks are the veins that let that economy thrive. Banks are the global network of connections circulating money across countries and continents. Banks set up international schemes and relationships to make transfers and payments with almost everyone, everywhere. They play a vital function in the economy and society as a whole.

In their role as financial intermediators, they transform short-term deposits into long-term loans. They allocate available funding at essential investments to stimulate entrepreneurship, health care, utilities, and much more. They allow us to save for our retirement and to sign up for a mortgage loan. Their insurance products make sure that you don't end up with completely empty hands when that house burns down. Their payment infrastructures allow us to send money everywhere across the world. One credit card allows us to pay in stores everywhere, making us the global citizens that we are today.

Banks are powerful organisations, managing and distributing money, a source of welfare and evil. Most banks will take good care of that and will make sure they are good custodians of their clients' savings, but exceptions make abuse of that power. Those exceptions are part of the problem of why banks today are often considered villainous participants in our society despite the fundamental role they play in lifting the world we live in, as I just explained.

When talking to bankers, though, I hear a different sentiment. When I ask them their purpose, common feedback is that they are there to help people realise their dreams. They allow people to own the house of their dreams and to have a safe place with their family. They bring belonging and a sense of community doing so. Those noble words come from the mouths and minds of C-level bankers.

Banks support people in financial stress with a consumer loan, and if you're lucky with your bank relationship, they may even have digital solutions to help you overcome scenarios of financial stress with personal finance management tools, etc. Banks see themselves as partners for new entrepreneurs, providing whatever needed to manage their finances and even a loan if needed. Banks make sure these businesses can trade and do business worldwide thanks to credit insurance and trade finance solutions.

Banks are there to realise dreams. But why are people so suspicious of hearing banks share this purpose? Is that suspicion justified, and if not, where does it come from?

Where does the divide in perception come from, and what can be done to align that perception again, making banking the trusted organisations that they claim to be?

"The New Ethos in Banking" was the theme of my events in 2022, after talking about "Trust" in 2021, and what a theme it was. I received endless opinions from people in the industry, sharing what it means to them. On stage, I had multiple CEOs join and explain how they look at ethos in banking; I had even more people interested in sharing their points of view in informal talks and more formal interviews.

All those voices made me realise that it is time for banks to embrace ethics and values for a meaningful transformation. As financial intermediaries, banks have an incredible impact on our economy and, as such, a huge responsibility towards society.

That means that banking, and many individual banks, must rethink its ethos.

We had executives explain the challenge of not leaving anyone behind in the digital age. Even in developed countries, people suffer from a lack of access to digital or a good understanding of how to use digital channels. With bank branches closing because of diminishing traffic, digital exclusion is also a banking problem. How big is that problem, and what is the financial industry's responsibility in creating a more inclusive digital economy?

Global warming affects everyone on Earth today. The Paris Agreement was the start of a set of measures to keep global warming under control in the long term, aiming to "keep the increase of the global average temperature to well below 2 °C above pre-industrial levels" and pursue efforts "to limit the temperature increase to 1.5 °C above pre-industrial levels" (UNFCCC, 2024). This resulted in many follow-up initiatives worldwide that also affect the financial industry. Some see it as another set of required regulatory measures; others see it as the biggest opportunity for the industry since the Industrial Revolution.

How does your organisation take it? And how will you ensure that you can take all your clients on that green transition? How will you ensure your clients understand what is at stake and help them choose their investments wisely?

This global challenge requires everyone to take action, but not everyone has the same means to do so. What can banks do to reduce or eliminate that inequality, ensuring climate action is not just a thing of the happy few? How will your organisation redefine its credit policy so that everyone can contribute, and can afford the investments required to future-proof their houses?

For centuries, financial services have been developed by mostly wealthier white men, and this persona has also guided product and service development in the industry. Unfortunately, some of that legacy is still part of the DNA in finance today (Broløs & Taylor, Designing for people—Learning from inclusion initiatives, 2024). Luckily, that is changing, and banks actively seek more diversity in their human workforce to ensure a good representation of society within their organisation. Is that enough to make banking services equal and appealing to all? Can more be done?

The world is changing, and so is technology. Over the past 25 years, we have seen an industry transform from a branch-heavy environment to a digital-only environment. The focus was efficiency and convenience for both clients and employees. The recent developments in Generative AI set new standards to what can be expected from a digital transformation. Although today, the focus is often more on threats than opportunities, I believe that in the near future, we will look at a much more effective digital environment, where efficiency and convenience will be equally important to empathy and personal care of both the customer and the employee. This will radically shift how we will look at technology and the interplay with people. It will redefine people's expectations of digital experiences and mark a new era of understanding of customers.

I am writing this book for precisely that reason. These societal challenges should also be the bank's challenges. If money makes the world go round, then banks are the veins that let the money flow. Banks determine the direction and speed of that flow.

To get a better grip on the flow, banks have focused on digital transformation for the past decades, and as much as that was essential for survival in the past, digital transformation remains a means to an end, as you will read later in this book. The emergence of new fintech organisations triggered a movement for a better world as they sought a share of the financial market. In response, banks quickly adopted new technology to maintain their leading position in helping people achieve their dreams. This digital transformation continues to evolve and attracts numerous service providers from the fintech industry.

Digital transformation has made banking more efficient, which is critical to coping with competition at competitive pricing. It also made banking more convenient, which is equally critical to fulfilling customers' expectations and, again, coping with competition from new banks, niche fintech players, or technologically savvy incumbent banks.

Digital transformation was an essential survival strategy, and it is still an important one, but today, we need to rethink banking at large. Banks have needed digital transformation to remedy pain points in their way of working. Digital transformation was the bandage on a wound. Today, however, technology is an essential part of banks. And now that the scars have healed, it is time to think beyond digital. Technology has evolved from features to core and will serve a bigger picture tomorrow.

Banking is at a crossroads of reinvention, and that requires another approach to transformation. That is why it is so important for you to read this book, not only to understand why banks often keep struggling to be the trusted financial advisor they claim to be but also to think about what role your bank should play and what is needed to do so.

The book is structured based on a couple of building blocks.

We will start with a brief history lesson on the evolution of banking. Banks have played a crucial role over the centuries, and understanding this history will help us reshape solutions for the industry to regain trust as partners in society.

The recent shift from branch banking to digital banking deserves a separate chapter, as it has completely transformed the relationships between banks and their clients. Despite being closer to their clients than ever before, many clients feel underserved. We will explore the origins of this sentiment and how to address it.

The New Ethos in Banking begins with a thorough understanding of ethos and its relationship to ethics. We will explain how ethos, pathos, and logos are interconnected as part of the rhetoric triangle. Chapter 5 will provide practical examples to illustrate the challenges of defining a new ethos.

In Chapter 6, we will explore how regulation impacts ethos and why regulation alone is not the solution. While regulations establish the rules of the game, they can also hinder banks' initiatives to benefit people and society. We will explore the reasons behind this and emphasise that regulation is not the sole answer but rather sets the direction.

Chapter 7 will outline a vision for the next wave of digital transformation, focusing on placing customers and employees at the heart of the transformation. Generative AI will enable banks to rethink their digital transformation, with their own ethical and value framework guiding the development, thus distinguishing one bank from another beyond just pricing.

Chapter 8 will examine the extent to which the Green Transition could be the most significant opportunity for banks since the Industrial Revolution. This transition requires massive investments, which can only be facilitated through effective financial intermediation. Banks will play a pivotal role in ensuring that deposits flow into green investments, similar to their role in accelerating the Industrial Revolution. However, as detailed in this chapter, there is much more to be done.

We will conclude the Ethos story with considerations for every institution to redefine its Ethos. Can banks regain trust as partners of their clients, architects of dreams, and financial doctors improving people's lives? We will explore what is needed to change the industry's perception and how banks can realign ethos, logos, and pathos to ensure their messages are not only understood but also accepted.

The conclusion of the book will not be a typical list of things to remember, but rather a future vision of banking. It will offer an updated version of a chapter requested by Chris Skinner in 2020 for a book looking at banking when my first son would turn 20 in 2040. This future vision will encapsulate key learnings and translate them into a possible future of banking, providing a fitting close to this book.

2

Once Upon a Time: An Historical Perspective on Banking and Trust

2.1 Introduction

I remember a conversation with a relative, an entrepreneur backed by a father with a good understanding of finance, who confessed that she is keeping her savings at home. I was flabbergasted. Why would someone keep their money at home in a country where less than 1% is unbanked (Lozano, 2022)? It's not that she doesn't have a bank relation. She has a couple of investments at a bank, but the savings stay at home. She knows that rationally speaking, it doesn't make any sense and that even a bit of interest is better than inflation eating the money at home, but that is where the savings stay for now.

That is not because the bank doesn't want to serve that person or because she doesn't understand the system. No, it is because she is fed up—this person's words, not mine—fed up contributing to the banks' profits given the little earnings it gets on the savings account (Coeckelbergs, Setting the scene of ethos in banking, 2023).

That same euro saved at home will also lose value as inflation never sleeps. So, distrust in an industry has a multitude of negative effects on the realisation of people's dreams, the overall growth of an economy, and our social construct at large. After all, every euro saved at home is a euro that cannot be invested in a mortgage.[1]

If you work in a bank, you must know what I am talking about when I quote a CEO of a bank saying, "It was difficult to go to parties. As a banker, I was bombarded with unflattering remarks. That was hard" (Vonck, 2024). I'm not a bank CEO, but I get the comment too. People trust banks with their money, but do they trust banks as institutions?

R. Coeckelbergs, *A New Ethos in Banking*,
https://doi.org/10.1007/978-3-031-78234-3_2

These two stories are symptomatic of the industry's reputation. They are counterproductive to the industry's mission and role of lifting society and its people to new levels and to the realisation of dreams.

Where does this distrust come from? Is it unique to today's situation, or is it something of all times?

In this chapter, we will examine the history of banking, the role that banks played, and what they meant to others' perceptions of them. Financial services are incredibly complex, and only a few people desire to understand them properly. People love what a bank can do to them, but most prefer avoiding the complexity behind financial products and services. How has this complexity grown over the centuries, and how has that affected the trust people have in the financial services industry?

Once we have a clear understanding of these fundamental questions, we can delve deeper. What lessons can we learn, and what lessons should we unlearn as we define the new ethos in banking?

2.2 How It All Started

Modern banking as we know it may be relatively recent, with the Medici Bank often referred to as the first modern bank in history, back in 1397 (Dinesen, Absent management in banking—How banks fail and cause financial crisis, 2020), but the foundations go back millennia. The grain banks in Assyrian and Mesopotamian kingdoms are often referred to as the first bank-like institutions dealing with foreign exchange and unsecured lending. The five-thousand-old cuneiform tables are the first proof of a banking model. Birch explains that four thousand years ago, the temples of Babylon accepted deposits and made loans with these deposits, and all that was documented on clay tablets (Birch, Before Babylon, Beyond Bitcoin—From money that we understand to money that understands us, 2017).

Throughout the millennia and centuries, banks remained important in facilitating trade and playing their role as financial intermediaries, but this came with risks since regulation wasn't the way it is today. Banks were local institutions that spoke to local institutions in other geographies. They were based on a common trust, not because of high levels of regulation but because of a consensus between the financial intermediary and its customers. People knew each other, and breaking agreements had immediate consequences.

Founded in 1397, the Medici Bank from Florence was likely the first true international bank. For one hundred years, this family business thrived, with ten branches across Europe. The Medici Bank's growth was mainly based on a good network of family members who each had their own branch and market to serve. International banking wasn't seen before and came with new challenges (Dinesen, Absent management in banking—How banks fail and cause financial crisis, 2020). Communication was mostly by horse, as these people didn't have internet, phone, fax, or even train yet. The more actual information there was available, the more money could be made. Asymmetric information made the bank smarter than the rest, and a reduced information asymmetry was part of what people paid for. In that respect, it isn't much different from today, with dealing rooms looking for the best location and connection for a few milliseconds of competitive advantage from their peers.

A combination of a worsening economic context, a tough political landscape, and a leadership that couldn't redirect the bank's strategic direction led to the fall of the Medici Bank in 1494. The family remained highly influential, though, with many nobles like Dukes, Queens, Cardinals, and even a pope (Clement VII) (Dinesen, Absent management in banking—How banks fail and cause financial crisis, 2020). In that respect, the Medici Bank was also an example of the status bankers get, even after misbehaving.

The key lesson, though, is that even 600 years ago, banking was more than just transaction management. Even then, the Medici Bank stood out because of information management, being the first to hear relevant news that could influence the value of the money it managed. Another lesson is that banking is a complex business that requires a constant compass on the world around it.

For this reason, this is often considered the first bank. With De Medici, banking became more than managing transactions and keeping money safe; they got into the information business on a much larger scale. Banks became much more complex, with international networks to manage both money and information that could affect the value of that money.

Bankers became risk managers and started to monetise information asymmetry to better manage money and money flows. Perhaps even more important, banking became a job that is hard to explain, making trust very important, as clients put their wealth in their hands with the promise to keep it safe. It was also an era where only the privileged had access to banking, and it created a powerful network of people that remained influential for generations to come.

Over the years, banking changed fundamentally, for good and for bad. So did the perception of the bank profession. A banker once recommended I read "The Value of Everything—Making and Taking in The Global Economy" by Mariana Mazzucato. The book describes how values and the definition of value changed throughout history and how economists defined models to calculate value. The book also details what were perceived to be value-creating roles versus value-extracting roles, productive vs non-productive classes and how that changed over time.

Interestingly, for a long time, banking was seen as a value-extracting role. Banks transferred value and asked for a fee for their services. As they only transferred money from one party to another, from deposits into credits, and applied remuneration for their services, they were perceived as a distributor of production, not a creator of wealth. That wasn't always appreciated that much (Mazzucato, The value of everything—Making and taking in the global economy, 2019, pp. 22–74).

Do you see the parallel with my relative here? She believes that banks return profits to their shareholders. In a low-interest rate environment, this leaves the impression that banks extract money from the market to allocate it to a much smaller list of shareholders, thereby forgetting the value of having money safe and even getting at least a small interest on it.

With waves, banks were seen as entities that extracted value from an economy, and later, they were appraised again for their distributive role (Mazzucato, The value of everything—Making and taking in the global economy, 2019, pp. 22–74). The Industrial Revolution was a unique opportunity for banks to lift society to new levels. It was time for massive investments to be made to set up a railroad network and electricity, steamboats and telegraphs, factories, and more factories. Governments couldn't do this independently and needed funding to keep or grow their competitive advantage. Banks were uniquely placed to accommodate that funding.

Banks made sure the required funds for productive investments were there. They promoted efficient usage of those funds, and through co-creation with the industry, they ensured the proper rate of return to grow further the interest to push for more investments (Da Rin & Helman, 2002).

Investment banks were at the forefront of this migration to the industrial age. New banks entered the market all over the world. They focused on large tickets and served big corporations and nations with money from financial markets rather than retail banking. This period marked the start of J. P. Morgan entering the market in the United States, quickly becoming the primary financier of the United States railroads, United States Steel, General Electric, and AT&T (Dinesen, Absent management in banking—How banks fail and cause financial crisis, 2020). Also, Europe counts many examples of new banks that filled the financing gap to help entire nations in the development towards the Industrial Revolution.[2] They all had one common denomination: they were not retail or savings banking. The regulator steered them mainly into buying government bonds because these savings banks got their deposits from households and "often lost them to fraudulent or excessively risky money-making schemes" (Mazzucato, The value of everything—Making and taking in the global economy, 2019, pp. 22–74).

Banks were the engine for a new world. They heated the industry economy, allowing enormous investments that sparked a transformation that the world hasn't seen since.[3] Banks were loved for this by politicians and society at large, as they reaped the benefits of banking financial disintermediation.

This positive sentiment towards banks declined again throughout the twentieth century. Their contributions to transforming society for the industrial age reduced while profits increased, making economists at the time believe that monopolistic powers were in play. With every bankruptcy of a financial institution or a new fraudulent investment scheme, consumers would lose their deposits. That influenced politics for more regulation, and the relationships with banks cooled down because of a distrust in the industry (Mazzucato, The value of everything—Making and taking in the global economy, 2019, pp. 110–112).

The 1930s and its global depression started a wave of new regulations. The Glass-Steagall Act separated commercial and investment banks, cutting down some of the complexities of the industry. The Bretton Woods Agreement and the gold standard that followed 11 years later further tightened the bank's playground and increased the transparency of international money flows[4] (Dinesen, Absent management in banking—How banks fail and cause financial crisis, 2020). Banks knew the rules of the game and did their share in the development of society, mostly without excessive risk-taking or complicated products and with good governance structures.

Throughout history, banking has been valued during times when it has been able to demonstrate a positive contribution to the world around it. However, when that contribution faded away, and the perception grew that banks worked for the shareholders only, or even worse, when they began to abuse their power and exploit information imbalances, they quickly came under scrutiny.

Banks operated in their own ivory tower, and the people and organisations around them failed to see how this contributed to a better world, if they already did, as most banks only served the privileged, business, and public institutions in the far past.

2.3 How the Freedom Era Led to a Wave of Deregulation

Until the early 1970s, banks were heavily regulated and perceived as value extractors instead of value creators. Most governments saw heavy regulation as key to controlling the market and avoiding financial crashes and scandals from the past. After decades of strict regulation, the pendulum shifted in the other direction starting in the 1960s.

Suddenly, regulation was increasingly believed to hinder the growth and development of the capitalist world, including the financial world. Countries, especially financial centres like New York and London, started competing to be the world's biggest financial centres, and the financial services industry became a highly competitive space.

The end of the Bretton Woods system and the economic uncertainty that followed, with zero growth and high inflation, were blessings for financial services. Banks saw an opportunity and took it. Instead of the end of capitalism, new transformations, and deregulation they led to a much higher market-driven economy. As the world continued to embrace the free market, governments got in an arms race to the bottom in reducing regulation. The deregulation opened new doors to product innovation and financial engineering (Mazzucato, The value of everything—Making and taking in the global economy, 2019, pp. 22–74).

There was one exception in deregulation worth noting in the 1980s: money laundering (indeed, that's how long it took to make it illegal). The free market was a blessing for many businesses, including the cocaine business. The outflow of dollars resulting from the cocaine shipping triggered Congress to pass a Money Laundering Control Act in 1986, making money laundering a federal offence in the United States (Alessa.com, 2023). In the coming decades, Know Your Customer, Anti-Money Laundering, and Anti-Terrorist Financing became increasingly important for banks, who were expected to be the gatekeepers against these kinds of illicit activities.

The increasing complexity and interconnectedness of transaction services would make it close to impossible to close the gates against criminal money, leading to billions of penalties per year and a damaging effect on the reputation of banking.

As capitalism evolved to become increasingly more sales-driven, so did financial services. Consumption was the oil to keep the economies growing, and loans were welcome facilitators to accelerate it. Banks were playing the game that governments wanted them to play.

The construction of ever-more complex financial products secured the values of credit portfolios. Mathematically, it all made sense, but it became nearly impossible to transparently evaluate the real assets behind these financial products. The industry believed it had risks under control but sometimes forgot that "zero risk" and "close-to-zero risk" are not the same.[5] As long as the underlying value reflects the credit, the industry didn't see a problem. But the moment this gets stretched, it is time to watch out, as the financial crisis has shown.

Customers weren't looking for complexity or risky loans, but they liked the promised return. They didn't understand how banks managed their money, and they didn't care as long as banks helped them realise their dreams. They became detached from their money and where or how their money was invested. The products were too complex to understand anyway. As people relied increasingly more on their banks to cope with that complexity, trust was vital for banks to survive.

Banking became more accessible, allowing more people to apply for loans, open bank accounts, and save money. The banking industry's influence on society and people's lives grew, but their sense of responsibility didn't always follow at the same speed. They were also given more freedom to innovate and conduct business. However, financial education was not always prioritised. People benefited from the free market, whether they had the funds to pay for it, or borrowed it from others.

2.4 The Rainbow Pattern Turned to Brown

Until the end of the '80s, banking was a relatively coloured industry. Most banks were small, local players with a straightforward business model, often created by founders who witnessed a specific need in their environment or community and set up the business at the service of that community. Others, like J. P. Morgan, remained global universal banks, building on the business they made in the Industrial Revolution. They were founded by smart entrepreneurs who saw a financial need and built a business around it.

Different business models coexisted and challenged each other. Depending on the region or country, you would have mutual banks, cooperatives, credit institutions, building societies, savings banks, and one or two universal banks. This coloured banking market sparked competition and provided a higher diversity of choices to consumers. Some were perfectly happy with the shareholder bank, while others preferred a cooperative or mutual model and participated in the capital. Excessive growth was impossible in a heavily regulated environment.

That changed in the 1970s and 1980s and accelerated after the fall of the Wall when the sentiment of victorious capitalism grew. Economic theory stimulated the overall belief that scale is the north star for efficiency gains across all industries. More volumes would reduce the cost to serve and increase shareholder value. On top of that, Europe was preparing for the euro, which was believed to start an evolution to a uniform market where only a few banks could survive. It led to a wave of consolidation. Banks would build a unified platform to service multiple markets. Slowly, the rainbow colours started moulting.

In nature, when the vibrant colours of the rainbow blend together, they create a brilliant white hue, with a fabled pot of gold awaiting at the rainbow's end. Sunlight, a mesmerising interplay of varied coloured light, embodies this phenomenon. This differs when you artificially mix the same colours in paint. Mixing all these paint colours leaves you with a murky brown, a much darker combination. This analogy mirrors what happened in banking.

As part of deregulation and the idealisation of the free market in the'90s, the financial services industry experienced a significant shift from being diverse to becoming more standardised. The shareholder model was told to be more efficient, with faster decision lines and less communication with the members. Llewellyn et al. explained how in the UK, many mutual insurers, building societies, and other cooperative financial institutions were reorganised into shareholder-model banks instead of the traditional membership structure, where profits are reinvested or distributed among the members.

In the free market, shareholder value was believed to generate customer value. The idea was that if management prioritised shareholder value, it would eventually lead to better customer treatment. After all, to maintain profitability, banks would have to treat their customers well. If customers were not satisfied, they would switch to other banks, causing a decline in profitability and ultimately reducing shareholder value.

I recently spoke with a former executive of one of the big strategic consultancy firms, who admitted to having been in the driver's seat of that transformation and explained that given their knowledge and existing beliefs in business, this evolution was the right thing to do. However, looking back today, he also confessed that this lack of diversification resulted in the banking sector's deprivation.

Short follow-ups on financial results would create transparency and keep management accountable for their actions. Larger-scale businesses could reduce the cost of banking to customers; that was the theory. This led to a market where most banks suddenly lived to the same principles and targets, which hampered their customers' options, resulting in lower customer satisfaction. The lack of diversity and competition also slowed down innovation (Llewellyn et al., 2014).

In 2011, Andrew Bailey of the Bank of England shared the same message, confessing that the demutualisation was a mistake. More diversification in bank models would have increased competition, reduced returns, and expected returns from shareholder banks. Banks have a crucial role in society, but reality shows that the financial industry at large was principally serving itself rather than society.

As the sector became more consolidated, competition diminished but internal complexity increased. While larger banks were expected to achieve better financial and strategic outcomes, they also grappled with integrating diverse cultures and organisations. This lack of a unified direction made managing these institutions more challenging, particularly as they inherited risks and legacies from each merger and acquisition.

For many, it also led to a detachment from the local communities and clients. Increasingly more, banks were being run from their "Ivory Tower", believing they knew the customers, based on assumptions rather than solid research and understanding.

Corin Millais, Head of Socially Responsible Banking at Teachers Mutual Bank in Australia, raised the same conclusion by saying: "Culture defines behaviour and actions, and we've seen what culture does. It's been a global phenomenon of unparalleled greed and growth with a byzantine world of complex structured securities, sub-prime, and in the end, assets and liabilities that simply could not be traded, and that, of course, was the Global Financial Crisis. Banking was so far away from people and communities that we all paid the price" (Coeckelbergs, About David, Goliath and The New Ethos in Banking, 2023).

2.5 Complexity as a Financial Product

The financial services industry was a traditional sector that relied on the same products and services for centuries. The sector grew at the pace of the market it served and its profits as well. That changed over. Deregulation and liberalisation, thorough evolution in capitalist thinking, the desire for growing profits and digital transformation made banks creative. Financial institutions found creative ways to innovate products and services to buy themselves in new markets like private equity, insurance, and investment products, thus developing a new market potential that didn't exist before.

Unfortunately, the rapid pace at which this happened was too fast. New product development was delegated to experts, but the management overseeing the bigger picture was no longer always capable of properly evaluating the risks and uncertainties associated with these new constructions. Additionally, all this occurred during a period of massive consolidation and industrialisation, adding many more challenges.[6]

The final removal of restrictions of the Glass-Steagall Act in 1999, with the passage of the Financial Services Modernisations Act of 1999, put the door wide open for banks to combine deposit-taking and investment banking, for some, a healthy strategy of diversification, and for others, a toxic mix that was waiting to go wrong.

Mortgages and other loans were repackaged into new financial products and resold to investors with a margin on the sale. Suddenly, banks could lend more, as they divested the mortgage risks by selling the packages. It was the start of much more creative product innovation, which would lead us too far for the sake of this book. This changed the game, leading many years later to the Global Financial Crisis. Mortgage brokers, the ones selling the loans, got an incentive often linked to the level of interest rate paid. That meant that sub-prime mortgages with a lower risk were more lucrative than prime mortgages. Mortgages were often structured so that the repayments were low in the first few years and went steep up afterwards. In case of financial problems, they could still sell the house.

A fear of missing out stimulated bankers worldwide to invest in these so-called low-risk, high-return investment vehicles, making them all dependent on the health of the American housing and mortgage market. Others confidently invested in these products, so what's the harm?

For banks, that was no problem in a market with rising house prices, where the underlying assets of a credit become more valuable, that way reducing the risk of mortgages over time. But what happens when prices go down, and people are unable to pay back their mortgages? And how do you deal with that if 25% of the home loans on your balance sheet are subprime? To make it more complex, how do you deal with packages of those loans sold to you as an investment? Credit rating agencies rated these mortgage-back securities but wrongfully assumed that with sufficient diversification, the risks would be well covered. After all, what are the odds that all these assets would default simultaneously?

The first concrete sign of a Global Financial Crisis was HSBC's profit warning in February 2007, saying it needed to make a provision of $11 billion for sub-prime loans, with chances that more provisions may be needed in the coming years (Dinesen, Absent management in banking—How banks fail and cause financial crisis, 2020). The rest is history. The industry was a perfect storm of complexity, leading to unfortunate events that still explain some of today's perceptions of banks.

New and more complex products come with new risks, which became strikingly clear with the global financial crisis. I spoke to people who explained the crisis as the result of changing accounting rules. However, most executives admitted that the Global Financial Crisis was much more than that. Deregulation and the fact that regulators always run behind the facts in innovation and new product development also played a role. The impact of the crisis required introspection as well. Why did these new complex products destroy so much value across the world? What went wrong in the risk management of these products, and how could it be avoided in the future?

One obvious conclusion was that greed and a short-term focus on profits were two more reasons for the crisis. My thesis is that banks got lost in their own complexity. The products got more complex, and the risks to manage were new and very hard to manage as well. On top of that, there was the human factor and a search for a uniform direction in many organisations that lacked proper governance after a wave of consolidation in the market.

I may be simplifying things a bit, but an industry-wide consensus that products were low-risk often made these products low-risk (and highly uncertain) until the market got into a cocktail of events that imploded these new products in ways that risk managers never expected.

2.6 Conclusion

Banking didn't evolve much for a very long time. Banks ensured that money flowed across the economy and that communities could thrive with basic financial services. Because of this, they were often seen as value extractors from the economy, with the exception of the Industrial Revolution. The industry experienced relatively normal growth, with a small crisis every after economic imbalance, in a world of uncertainties.

However, deregulation since the 1970s triggered a tremendous change. New technologies and liberalisation brought new opportunities for the industry, which led to a speed that was hard to manage. Banks got in a storm, and instead of mitigating the reputational damage of misbehaviour, the industry was driven by greed, focusing on shareholder value and growth. I'm explicitly stating the industry because, of course, not all banks were bad, and likely only the minority misbehaved, some small, but some very big organisations.

Banks created new products and unknown internal and external dependencies, the impact of which was impossible to predict. The fear of missing out overtook the need to understand, and new innovative financial products took the world at a speed never seen before, thanks to globalisation and consolidation of the industry. Envy, followed by greed, complexity, and indifference, reinforced each other into a poisonous mixture.

With every misstep come penalties and bad press. Every crisis brews new regulations to ensure the industry does better going forward until the pendulum goes into deregulation again, under the false belief the industry has everything under control. Today, banks have higher capital requirements and a better resilience framework, so a crisis like in 2008 is unlikely to happen again. The recent collapse of Silicon Valley Bank has unfortunately shown how nervous the industry still gets when facing a bank in trouble. That nervousness shows how fragile the industry still is. The moment banks lose trust in each other, liquidity runs dry, solvability comes at risk, and a new crisis looms around the corner in a matter of days.

Over the past 50 years, banking has become very complex, with interdependencies across all service lines, both on an organisational, human, product, and technology aspect. The wave of consolidation brought a mix of cultures that were sometimes difficult to manage and tough to control, adding to the already high levels of complexity. Complexity that is hard to manage leads to trouble and to regulation.

Regulation is a sign of distrust: rules are made to ensure everyone follows the right path when there is "a reasonable chance some may deviate from that path for their own benefit".

One last important additional element of complexity is how the digital transformation of the past 20–50 years transformed the industry and affected the way people look and engage with banks. In the next chapter, we'll look at how the industry embraced technology and what it meant to its identity perception and its mission of realising people's dreams.

All of that should offer you a basic understanding of where we are today, why a new ethos in banking is needed, and what potential steps are to a new ethos in banking, one where values and ethics are embraced for a meaningful transformation.

Notes

1. In reality every euro not put on a saving account equals a multiple of money not issued as a credit (Manning, 2022). Suppose banks need to keep 20% of their deposits as a required minimum at the central bank, in theory every 1€, or dollar for the matter, can make 5€ or dollar of credit creation.
2. In Belgium, the first country to follow Britain in the Industrial Revolution, Société générale pour favoriser l'industrie nationale was the world's first joint-stock investment bank, created in 1822, and an active contributor in industrial finance from the early 1830s. With Banque de Belgique, they capitalised about two-thirds of total industrial credit banks in Germany, it was Kreditbanken that took the lead in industrial finance, and more specifically Schaaffhausen Bankverein, Disconto Gesllschaft, Bank für Handel, und Industrie and Berliner Handelsgesllschat (Marco Da Rin, 2002).
3. I see you think: what about the Internet? Maybe, although the real transformation of the connected society remains yet to come.
4. The Bretton Woods system fixed the USD and fold at one ounce of gold to $35 and with the USD being fully convertible into gold. Members had a fixed parity with the USD directly and indirectly with gold. The control mechanisms on international trade flows between 1944 and 1959 limited the complexity of bank activities.
5. Extreme risks were being repackaged and sold again, universally accepted and reinforced by the industry as reliable new banking products. Mortgage-backed securities (MBS) were invented in 1968, but it took until the '90 s before they reached a critical mass. In 1996, the volume of MBS was less than $500 million, but it more than doubled in two years to $1 trillion. Five years later, in 2003, the total issuance of MBS was $3,2 trillion (McConnell & Buser, 2011). Business was booming, banks realised people's dreams, as credit risks were being minimised, allowing more people to get loans or credit cards, assuming those risks could be repackaged, 'securitised', and resold to other investors.
6. A milestone in that respect was the Depository Institutions Deregulatory and Monetary Control Act of 1980, which removed, for example, the limit on interest that banks could pay depositors. The limit stimulated consumers to look for better rates at unregulated entities until then. Because of that, the assets under management at banks were sufficient to serve the best clients with low credit risk, but not more.

Removing the interest limit allowed banks to attract more deposits, nurturing the need to grow the production of credit and mortgages. This triggered banks to deal with riskier clients and more lucrative interest rates, known as the subprime market.

3

How Digital Transformation Ended Up Being Much Bigger

3.1 Introduction

Over the past centuries, the banking industry has undergone an enormous transformation. With society, banking modernised into its current self, with ups and downs. The changes over the past decades were never seen before. Emerging technological advancements, more sophisticated products, and the hunger for growth and economies of scale all contributed to a much more complex industry. It made banks more challenging to manage and necessitated new rules and procedures to handle the associated risks.

From the earliest days of mainframe computers to the introduction of magnetic stripes or cash machines and payment cards, banks have always been at the forefront of technological innovation. They have continuously strived to stay ahead of the curve, adopting innovations that improve efficiency and foster customer loyalty. It led to a rise in micromanagement, on the one hand, with experts driving the change but sometimes forgetting to connect the dots in a web of interconnected causes and effects.

Digital connectivity has become an integral part of our daily lives and often manages our routines more than we manage them ourselves. Today, there are more smartphones in circulation than there are adult people across the globe. As a wise speaker, Roland van der Vorst, once said at The Banking Scene Conference 2024 Brussels, and wrote in his book: "We used to live and act with the end in mind, but the digital world has another logic: it keeps to moving to the next thing, over and over again" (van der Vorst, 2023). What does that mean to how people look at digital banking?

R. Coeckelbergs, *A New Ethos in Banking*,
https://doi.org/10.1007/978-3-031-78234-3_3

Over the past few decades, the banking industry has undergone significant changes, and the advancement of digital technology is the primary reason for it. Banks have built more efficient trade routes, making real-time transaction management possible. They offer more complex products thanks to digital connections that allow more complex risk management. With ever-growing data and fibre optics capable of transmitting data at the speed of light, sophisticated algorithms ensure that banks remain at the top of the food chain in terms of managing information asymmetry. Every nanosecond is a win.

Digital transformation made banks more efficient in managing products and customers. It helped them provide their services in a convenient manner, and it automated and scaled operations. Additionally, the digitalisation of daily transaction flows has created more transparency and faster money flow across the world.

We, users of digital services, are constantly adapting to the new reality that changes every single day. Digital services have added many additional layers of complexity, needed to reduce complexity for the end users. This means that these digital services play out very well for people who know how to work with digital channels. On the contrary, it brings new risks of exclusion for some people, and it can lead to sometimes unexpected abuse. The industry has moved from a few big releases per year when I started my career back in 2007 and evolved to incremental changes, sometimes up to a daily basis. It is a constant, careful balance of improving services in such a way that it doesn't feel much like a change. After all, who likes change?

Today, people need to trust an app or an online bank platform, whereas in the past, they trusted the advisor they spoke to. Personal contacts have become touch points, and the depth of customer relations is translated into one number: the NPS score.

The rise of digital transformation brought banks closer to their customers by increasing convenience and efficiency. On the contrary, it also removed much empathy from the process, a crucial factor in banking, an industry that remains highly dependent on trust and interpersonal connection, whether online or offline.

Therefore, a separate chapter should highlight the importance of digital transformation in shaping the new ethos of banking.

3.2 Banking Is a Technology Industry

It may surprise you, dear finance professional, but banking and finance have always been front runners in adapting new technologies. Electronic Funds Transfers date back to 1871, when Western Union started to make payments by telegraph. 34 years later, in 1914, the same company launched a charge card for their best customers for deferred payments (Birch, Before Babylon, Beyond Bitcoin—From money that we understand to money that understands us, 2017).

The origins of digital electronic payments can be traced back long before computers. Banks first started using the Charga-plate bookkeeping system in the 1930s. It was a metal plate with the owner's name and card number embossed on it. The customer could hand over the plate to the merchant, who could swipe it to transfer the cardholder information onto a sales slip for record-keeping, where the customer had to leave a signature to complete the purchase. The merchant would go with a copy of his bank to finalise the payment.

This was followed by the invention of the Charg-It system in 1946, which allowed people to buy from multiple nearby vendors using metal plates similar to the ones from the army. They could charge these purchases to the bank, without having to bring cash in the store. It was an early form of consumer credit, introduced by John Biggins of Flatbush National Bank (Girling, 2021).

It led the basis for how you and I pay today. This was the first concept where transactions were centralised and consolidated before being billed and reimbursed with an intermediary, in this case, Flatbush National Bank.

If you don't work in the payments industry (and even if you do), you may not know about these developments, but the next step in the evolution may sound more familiar. In 1950, Frank McNamara and his business partner rolled out the first credit card, the precursor of modern digital payments. It was the start of the Diners Club. Eight years later, BankAmerica issued the first revolving credit card and two years later, Forrest Perry invented the magnetic stripe at IBM, which became a global standard in the early '70s. This technology made credit cards much less fraud-sensitive; it didn't prevent fraud by skimming, though. For the innovation to prevent that, we must wait until 1984, when chip technology is added to the equation (Saks Frankel, 2021). It takes some time to make it a global standard, but today, we cannot imagine a card, whether that is debit or credit, to come without a chip.

Chip technology is as young as I am; I'm 40 at the time of writing. That's how fast the industry has evolved over the past 50 years.

We complain about the mainframe as being legacy old technology. I cannot blame you, at least not for the technology being old. IBM released its first mainframe computer in 1964, and it revolutionised everything, some sooner than later. Banks were at the forefront of these developments. The first ATM, the interpretation of magstripe data in credit cards or barcodes, but also banking on mobile phones would have been close to impossible without mainframe technology (Toole, 2014).

In 1985, the Bank of Scotland became the first bank to offer electronic home banking services to its clients. It took nine more years for the first website for bank services to launch. In 1994, Stanford Credit Union was the first to introduce this new service (UBS, 2022).

Since 2000, and maybe more since the financial crisis, the pace of change accelerated incredibly. The dot-com crisis filtered out good and promising technology from the noise, and the internet reached a level of maturity that allowed high-trust industries like financial institutions to transform their rules of engagement.

Until the Global Financial Crisis, smart entrepreneurs founded new ventures in the finance space, but they were not threatening, not yet. However, the dissatisfaction after the GFC and the democratisation of software accelerated the creation of new financial technology initiatives.

Clearly, banks didn't get the time to fully recover from the GFC before facing the challenges of a new revolution. The launch of the iPhone in June 2007, and competing brands with an Android operating system (of Windows for a short while), sparked a new era of digital transformation. Initially criticised for being too expensive, every year a new edition with a better customer experience and more possibilities was introduced. In combination with faster internet, the rise of smartphones kickstarted mobile banking: banking in everyone's hand, 24/7/365. We cannot imagine a world without it, and yet the technology is often not more than 15 years old.

Digital transformation revolutionised the entire industry, bringing banks closer to their customers than ever. However, in the process, the personal connection between banks and their clients faded away. How has this affected the relationship between banks and their customers? What impact has it had on the customers' perception of their bank? How do banks cope with the customers' changed expectations?

Every year brings a new series of buzzwords "banks must invest in". Wearables, RPA, blockchain, ICO, and, more recently, the metaverse, augmented, and virtual reality and AI are just a few of such themes.

Most banks were among the early adopters of digital technology compared to many other industries. It was, and sometimes still is, a revolution with much resistance, but the fear of missing out spurred the banks to their limits and pushed them to transform into a digital self. Efficiency gains and adapting to changing customer expectations were the main drivers of this transformation. "Because others do it as well" was a common internal driver for change, which is more out of personal experience and observation than from what I read in publications.

Recent research showed that between 2017 and 2021, the European banking industry accelerated its digital transformation, with a 15.82% increase in IT expenditures over total operating expenses, with the COVID-19 pandemic as a major driver. The study concluded: "What is always crucial is being committed to facing a technological revolution by trying to lead it instead of fighting or postponing it. In this case, it seems that the European banking industry has opted for being the protagonist of the digital transformation rather than being a mere spectator" (Carbo-Velverde et al., 2023).

Despite room for improvement, most of the interactions with a bank happen digitally, and if not, it is by the customer's own choice. For a long time, banks have been warned that neglecting digital transformation could lead to their demise. But, in reality, it seems that close to no banks have gone bankrupt due to neglecting digital transformation, despite the rise of neobank alternatives. In fact, I recently asked experts online to provide me with the name of a bank that went bankrupt because it missed the digitalisation train, but none of them could come up with an answer.

Of course, the evaluation is not that black and white: as much as unsuccessful digital transformations may not result in bankruptcies, they could make banks an easy target for acquisitions. After all: a bank that cannot efficiently work in a digital world today cannot properly work at all, which brings down the price.

Banking is about realising dreams, and today, that happens online.

3.3 The Fintech Revolution and the Quest for Good

Ever since the financial crisis, the world has looked at the rebels in the fintech industry as a source of good. People viewed fintech as a driver for change in finance, and rightfully so! This younger generation wasn't affected by old habits, money culture, and greed.

Whereas banking used to be a unique skill, digital banking found inspiration from different industries, technology players, and a new breed of entrepreneurs. The steep rise of business models and technological developments came right after the global financial crisis. Coincidentally, the enormous opportunity of having an app store in everyone's pocket emerged simultaneously as people started to understand the banking industry's misbehaviour before the crisis and the impact it had on society during and after the crisis. In that respect, the GFC was more than a financial crisis. It also sparked a digital revolution, as well as an ethical one.

That crisis has shown us that things have to change for the better. Young, bright minds stood up, raised their voices, and took action. The democratisation of software made it increasingly more convenient to build new digital solutions, whether that was to improve banks or to better serve customers themselves. They set up their own businesses to challenge the status quo in banking. Convinced they could do better, they raised the necessary capital and built digital financial services.

Initially, it was very much purpose-driven. People were frustrated by their existing financial services providers and found innovative new tools to overcome those frustrations, letting customers enjoy the efficiency gains with lower to no costs and often promising more ethical business models. They built new solutions because they saw a business opportunity by making things better, they wanted to improve financial exclusion or promote financial well-being.

Regulators set up new regulations, laws, and directives like the open banking initiatives to allow non-banks to apply for lighter licences, with limited regulatory requirements that allow new players to compete with banks, making the financial services industry more coloured again. It facilitated integrations and data sharing, with the ultimate mission to improve competitiveness. They learned from the 1990s and accepted that diversification of business models results in a healthier industry overall. Neobanks like N26 and Revolut set the road to improved customer experience and were an example for other banks to keep up.

People who didn't trust banks or were disappointed by the slow pace of their bank's digital adoption suddenly had an alternative.

Over time, lower interest rates and promising returns attracted more equity, which expanded the fintech revolution further. Fintech companies are known to be very powerful in very specific services. They provided lower-cost remittance services (like Wise) to ease the lives of migrant workers or peer-to-peer lending (like LendingClub) for people who didn't get access to the regular credit system or people in search of alternative investment options. Others would focus on merchant services and e-commerce (like Stripe and Adyen).

Because their strength was their focus, and their focus was on doing a very specific thing very well. Like you eat an elephant (or a very big pie): they ate banking piece by piece. Disrupting, and for some, breaking the banking industry was the ultimate mission. They built new exciting services or reinvented existing services to delight the customer. Banks started to see some of their revenues disappear, but more importantly, the connection with the client weakened.

Statista shared that fintech funding grew from an annual average of $6 billion between 2010 and 1012 to close to $63 billion a year between 2015 and 2017. 2018 saw an explosion of $148,6 billion, with a new high right before the pandemic in 2019 of $216,8 billion being invested in fintech. The record high until now was in 2021, with funding reaching $247,2 billion (Statista.com, 2023).

That means that between 2010 and 2022, a total of $1,2 trillion worth of investments were offered to these new tech companies to change the financial world. That reflects the opportunities that people saw to make money on the back of these fintech founders and also the number of founders that were capable of building compelling use cases worth the investment.

The fintech industry may not have broken down banking so far, but it did challenge them and forced them to transform and improve their services to become (more) digital. Some banks have developed their own offerings, while others have worked with fintech companies to keep pace. Over the years, the partnership model has become increasingly popular. Banks and fintech companies are now joining forces to find a win-win solution, with banks bringing regulatory expertise and clients, and fintech companies developing the tech stack and bringing creativity. This collaboration has proven to be the best of both worlds, although it remains a constant challenge to reduce friction from the process.

3.4 Efficient and Convenient Banking in a Digital World

Digital banking is the new normal these days. It was a long and sometimes troublesome process to get where we are today, and that should not come as a surprise since digital changed the face of banking. Much of the internal processes remained untouched for a long time. It is complicated to rethink processes. That is why it is much easier to think of improving existing processes instead of rethinking them. The same happened with the internet and digital banking. It started with the digitalisation of processes. People looked at how things worked and added another layer over it.

Lipstick on a pig.

It is helpful, though, especially in high-volume processes. Doing digital is a scale business. Once you set up the infrastructure to allow straight-through processing, most of the expenses are made and what is left, the maintenance cost will only marginally increase with higher volumes. Replacing people with machines is a tremendous cost saver for two reasons: most processes are cheaper and faster operated by a machine, and automation leads to fewer mistakes and better data integrity. Your digital transformation comes with a cost, but the mid-long-term gains make you a positive business case. Lipstick on a pig, but helpful.

The first digital transformation iterations focused on efficiency and convenience. Digital banking must make it easier to serve people and less expensive to do so. How do you take people on board in such a transformation? Because the main driver for a successful technology transformation is not the technology. We have every technology available to make whatever we like. Technology is not the problem.

The key success factor is the human element. It is the people who set the architecture, the people who build the tools, and even more, the people who need to use the tools. That is not my conclusion, but a key message at close to every event I have organised so far.

People don't like change, and banks experience that resistance day-to-day as they do digital. That also explains why digital change programmes do not always lead to instant cost savings. Whether it is customers or employees, people need guidance and understanding of why things change to embrace change. Good change management is golden and more important than the technological change itself.

Automating 80% of the processes will also make it much easier to calculate the cost of exceptions and make banks reluctant to deal with them. The result is that these exceptions would cost the customer much more in the hope of nudging them to change their behaviour to avoid the same exception in the future. So banks started charging more to clients who didn't ask for it on things that could have been prevented if the process had been bullet-proof in the first place.

If people don't understand the reason for the change, they may disconnect and feel underappreciated. That leaves scars on the relationship between banks and customers (and/or employees, depending on the affected process). Banks embraced technology for more efficiency and convenience, assuming easier-to-use bank services will convince people to use them, as happened in Asia, for example. In developed countries, banks often forgot that the update of fintech in Asia was because people went for nothing to something, whereas their clients in developed countries were pushed from branches and personal services to digital services. People interpreted digital banking and the reduction in branches as a cost exercise for the benefit of banks and their shareholders and missed the benefit for them.

Digital banks have become a more rational version of themselves. Banks were reconstructed with 0s and 1s, and the evaluation for success was done on a scale of 0 to 10. Everything needs to be scripted, or at least 80%. When we hit the 20% not scripted, we get lost and are unable to help the customers instantly, an expectation that the industry created by doing digital. We added another layer of complexity. Because things got so complex, we made everyone an expert. Everyone has their scripts to learn and master. And what happens if something goes wrong in between two scripts, two worlds that are blank to each other?

Banking became a set of building blocks, and each block has an expert. That logic is also translated into a bank's day-to-day human services. The time of one bank advisor to help you with your financial life is over. There are advisors for investments, others for lending and mortgages and often different ones depending on the loan you need. To make everything more efficient, retail customers no longer get their personal advisor, but they get an expert who is available, making sure everyone can serve as many people as possible, according to the scripts.

The more fundamental changes are, the harder it is to sell. Luckily, the healing process can be relatively short. People quickly understand that the efficiency or convenience gains also play in their own favour. We saw that in many countries during the pandemic with contactless payments. Using contactless was seen as a health measure at first. After a few taps, people started to appreciate the convenience. In Belgium, for example, 79% of the population used contactless minimum once in 2023, long after long after the initial safety concerns had arisen, compared to 47% before the pandemic (Febelfin, 2023).

"Because everything (in digital banking) is predetermined, everything is hyper-scripted, and there's only one outcome; it is meant to be efficient. And this is incredibly limiting. We think that efficiency is hyper-logical. That's not necessarily the case. I don't believe that efficiency is the most logical approach in part because it misses empathy and emotional intelligence. Why? Because the outcome needs to take into account the contextual whole", this was Ghela Boskovich's observation that she shared at our conference on November 20,

2023, themed "Augmented Empathy" (Coeckelbergs, When art and tech dance together—A review on augmented empathy talks, 2023).

Building a digital banking experience is an efficiency game. The business case is cost savings rather than customer acquisition or retention. This is a missed opportunity.

Ghela advocated for more effectiveness going forward, which will let banks develop their digital capabilities within an ethical framework with empathy for customers, the team, employees, and investors. Effectiveness involves considering the wishes, feelings, and preferences of stakeholders. This means that we need to approach bank products and services differently, as we will explore later in this book.

Let me give you one example, though. When I was looking for a mortgage back in 2019, the first option to get an appointment with an advisor in a branch was four weeks later, the alternative was the next week via a web call. The choice was a no-brainer. With some reluctance—I don't like change—I agreed. I wanted to buy a house, and every day I couldn't make a bid would increase the risk that they'd find another buyer. The web call happened to be a great experience, and delight took over from the feeling of reluctance. The time savings were enormous: I could have the appointment after office hours, and I didn't have to go to the branch, so an appointment of 45 minutes was effectively 45 minutes, instead of 1,5 hours lost for commuting and appointment. More importantly, I still had personal contact to guide me in a life-changing event: buying my own house. It was highly efficient and effective. I got a quick service, with a human being answering all my questions instantly and listening to my concerns and uncertainties.

In most banks, this remains a future scenario. Banks remain reluctant to interfere in classic processes for life-changing events. One of the speakers at the event on November 20, 2023, Bart Hamers, Director of AI and Analytics at BNP Paribas Fortis, rightfully noted: "The whole digital interaction that we have now has become very transactional and, therefore, a lot less empathetic". He agreed that new technological developments can help to bring empathy back into the equation and agreed that this needs to be done in an inclusive manner, including the entire society.

Efficiency gains have been the main driver so far. In developing the requirements, we kept the end result in mind and worked towards that with a balanced number of options to avoid uncertainties and risks of failure throughout the process.

Employees can be forced to join a change trajectory, as it is part of the job, but nudging is much more difficult for customers who are in the driver's seat for accepting a message or not. A bank can only interact whenever the customer is open to it. The customer decides which emails get in spam, or which notifications he allows. The customer decides which message in his inbox to open. That has created a distance between banks and their customers, making it much harder for banks to realise dreams.

3.5 (No Longer) Everybody's Friend

In today's world, we heavily rely on digital services. We shop online, watch online, connect online, and these days, we buy our cars online. Our world has changed, and so did consumer expectations. Services need to be available all the time and accessible from everywhere. 24/7, 360°, across the world.

Today, if there are any branches, they are at the service of digital platforms, not the other way around. Everything that is told in a branch, every customer contact or advice in a bank, and every deal that is signed in a branch is supported by a digital backbone. Even a neobank like Nickel whose USP is the physical network, would not survive without a solid digital platform that connects the physical touch points with a central IT system and mobile banking solution[1] (Nickel, 2024).

A study by the European Banking Federation showed that between 2008 and 2021, the number of branches declined from 273.723 to 138.376, −50% (Saravia, 2022). Nonetheless, branches remain incredibly valuable in some regions. We do see that the role of a branch has changed. Having a branch nearby creates bonding, makes the bank part of the community, and builds a personal relationship for the moments that matter to a customer.

The more commoditised services are, the more people expect them to work digitally, always, and instantly. Starting with daily banking services at the beginning of this transformation, a bank is expected to have all its products and services online, whether those are investment and savings products, lending products, or insurance. People with spare money want to pay with it or to put the money to work through savings or investment products, and they want that instantly, while others who need money quickly prefer to avoid the embarrassment of speaking to an advisor for a personal loan or wasting time in the mortgage process. Digital banking has become an integral part of daily life, whether one likes it or not.

The vast majority likes digital banking. The challenge for banks is that they serve an incredible number of unique people, each with their own preferences and expectations. In building digital solutions, banks looked for the common denominator to please as many people as possible. Personalisation was a hot topic, but it regularly didn't go much further than a general notification or a background image. True personalisation was left to a few digital frontrunners, and even there, reality proved much harder than the slide decks promised. Personalisation and hyper-personalisation services are not new but often don't fulfil their promise.

Banks that have pivoted to more delightful and personalised digital services are facing the challenge of capturing exactly what customers want and providing exactly that. Aside from the data challenges of capturing the right data and making the right conclusions, new technologies come with many more challenges.

Also, what technologies are worth investing in? How do you predict the success of new technologies? How do we decide whether to invest in new technologies if they have no proven track record yet? How do we keep the early adopters, the fast followers, and the average customer happy?

As I said, the vast majority likes digital banking. Certain clients will readily accept a bank's testing of new technologies, but a universal bank cannot afford to disregard more traditional clients who may be more hesitant to use digital services, whether that is because of an attitude or lack of knowledge using digital tools. How far do you go as a bank, where do you draw the limits, and how do you ensure to get as many users on board as possible to use new tools?

Internet banking took a while before everyone embraced it. Mobile banking went much faster. More people enjoyed financial services as the cost to serve over the internet declined, and new players like fintech and telco providers joined the market for financial services. Where access to financial services was limited to someone's assets under management and a credit score in the past, access to digital channels today is a more important factor for financial inclusion. Digital financial services were an accelerator for inclusion in regions like Asia and Africa because most people had a phone, and the phone connected them with others and with a bank account. In many developing regions, the phone enabled trade and business development. A mobile phone, whether smart or not, was a life changer for many.

The next wave was wearables. These were much less successful. Which investments bring delight, and which digital developments have a negative effect on a bank's image? Which services are considered a cost for the client, and which are of value?

And where lies the Holy Grail of innovations for which people like to pay more than they already pay today? Today, there are 7 billion smartphones in circulation, much more than the 2 billion computers. So clearly, mobile banking cannot be ignored. Compare that to the 250 million smartwatches in circulation, and you understand that the average retail consumer does not see sufficient value in wearables today (Caudron, 2023). Banks have invested in wearable devices with NFC technology, and some still do. I was part of a test group of KBC Bank in Belgium to evaluate the success. Some were offered a ring; others were given a keychain. I was offered a traditional watch with a chip in the leather bracelet. It was a fun gimmick for a while, and I know someone who could barely live without his payment ring. However, the moment that these devices were no longer subsidised by banks, the interest quickly dropped to the level at which banks stopped supporting this innovation. It didn't add enough value for people to pay for it, at least not in Belgium and not for me. My friend Andrew Vorster, on the other hand, would switch banks just to keep using his smart ring for any possible payment. How to satisfy every customer?

Clearly, mobile banking cannot be ignored, but even in that context, there are boundaries of what to do and what to ignore. That is very contextual. For example, my bank will send me a replay of the goals of my favourite Belgian football club in real-time. Personally, I see it as a great feature, but many clients didn't understand why the bank spent money to implement it in their app. Most other banks consider this as a step too far from their core business. These are strategic considerations that play a vital role in ethos banking, as we'll learn later.

Finding the right balance is a constant challenge for banks today.

Did you hear of "the rising tide from the retail side" before? A friend and awesome book reviewer referred to this concept to explain how our consumer experience also translates into what we expect in a professional context. We came from an era where professional environments were more advanced than our home environment. Transport, utilities, and the internet have found their way to the world through economic development and business. Today, that pivoted the other way around. Expectations changed based on our day-to-day behaviour. That means that like retail and corporate banking, private and wealth banking also needs to adapt to keep its clients happy. People built a framework of expectations on how to do things and how things should work, and that framework is mostly digital today. Retail banking, corporate banking, and even wealth banking simply cannot survive without an app, nor can any other type of bank without its digital platforms.

Customers expect digital banking. "Customers" is an average of all the individuals in society. But how about the ones that are left behind? Today, it no longer depends purely on a bank's willingness to serve but also on a consumer's access to digital services and understanding of both. In other words, someone may wish to invest an incredible amount of money, but without access to digital channels or the knowledge to manage those channels, that person is financially excluded today.

That proportion of people is larger than you may think. In 2022, Belgium saw an astonishing 46% of the population at risk of being digitally excluded. 46% of Belgians need to be more equipped to live a digital life, whether that is because of a lack of tools or the lack of understanding of how to work with them in a safe manner. That gap is growing. The segments most at risk in Belgium were singles (48%), people with a low income (55%), the elderly (60%) and people with lower education (69%). Ilse Mariën, Policy Advisor for digital inclusion at the Flemish government, explained it as "It is one thing to have the tools, another to have the so-called 'button knowledge', and another to know what is behind the use of those buttons" (Coeckelbergs, Digital inclusion in Belgium Anno 2022, 2022).

Giving people the tools is not a solution to include more people in a digital economy, to let them deal with financial services in a digital way; a free phone will not make people digitally included. Guiding people to the bank app will not make them digitally included. It requires training and understanding of how these phones and apps work, and it requires support in case they don't work.

In developed and developing countries, societal inclusion isn't possible without financial inclusion, which isn't possible without digital inclusion. Banks have a responsibility on all these levels, a responsibility they share with many other stakeholders involved, especially in a world where cash starts being marginalised.

Banking is about realising dreams, and that happens increasingly online. How will they allow every citizen a decent life in a way that makes them included in society?

3.6 Rivers of Data Fill the Lake of Opportunities

Open banking and the data exchange that followed (or didn't follow) is probably the biggest trend in fintech over the past decade. It was initiated in Europe by the Payment Services Directive 2 (PSD2), and in the UK by the Open Banking Regulation. Banks were compelled to open up and exchange their payment data to blow a new wave of innovative approaches to personal finance management, credit scoring, and many other areas. More regulations like the PSD3, the Payment Services Regulation (PSR),[2] and Financial Data Access Regulations (FIDA)[3] are pushing the industry to do more.

Open banking is not unique to Europe, by the way, many other regions followed. Some followed the same regulatory approach as in Europe. In other regions, open banking evolved with the market. China's Alipay is a great example of that. They were less pressured by mandatory requirements or regulatory restrictions. They simply built it across the world this was seen as one of the examples of the possibilities that open banking could mean in terms of innovation (Zhu, 2024).

The United States has had financial data exchange and payment initiation with third-party platforms for many years through web scraping, where a bot uses customer log-in details to access bank accounts. However, in October 2023, the Consumer Financial Protection Bureau (CFPB) proposed a Personal Financial Data Rights rule to shift towards open banking using more secure APIs instead of web scraping (CFPB, 2023), because web scraping is considered unreliable and risky. Open banking is said to achieve the same results, but in a much safer way, through the use of Application Programming Interfaces (APIs) (Taylor, 2024).

Banks didn't ask for it, and most considered it a pure compliance exercise in the beginning. They set up a minimal API platform on their own or found a partner to do it for them simply to comply. There was no incentive to do so, as the platform didn't make banking more convenient for them, nor efficient. They feared to be losing control over their clients. That fear was ungrounded. But open banking did trigger something. It changed the mindset and perception of payment data. After all, what is open to third parties is also open to banks. Creative minds started to look for innovative new business models, some more successful than others. Easier connections with accountancy software, better control over personal finance management, more valuable credit scoring, and more seamless top-up of accounts are often cited use cases. People started thinking out of the box to find new models based on the new mandatory platform to comply with PSD2, later PSD3, PSR, and FIDA.

That is essentially what open banking is to me: it is a mindset change. It inspires customers to view their banking services differently and, conversely, motivates banks to reinvent how to engage with clients to make the digital experience even more convenient and efficient.

PSD2 prepared banks for deeper integrations. It forced them to set up new infrastructures, which resulted in no revenues unless they found creative ways to build more value-added services beyond what was legally required in these new infrastructures. Open banking comes in many forms and shapes.

Open banking and PSD2 are often perceived as interchangeable concepts, but I disagree that they are the same. PSD2 regulations specifically require banks to establish an API marketplace for payment initiation and account information services. This means that banks are now obligated to grant third-party providers access to their customers' current accounts, allowing them to make payments or obtain information on behalf of the customers.

In my opinion, open banking is a much broader concept. Banks that embrace open banking to its full potential can redefine the rules of the game, setting clear approaches on what can and cannot be done and under which circumstances.

In Europe, the European Payments Council is trying this with the development of the SEPA Payment Account Access Scheme, a scheme "to drive 'open payments' in the European Union (EU) in a way that unlocks and creates value whilst allowing for a fair distribution of value and risk between scheme participants. The SPAA scheme covers messaging functionalities. It is not about a payment means or a payment instrument, but it offers a way to transport information in relation to payment accounts and transactions" (European Payments Council, sd).

Individual banks that develop embedded finance products are great examples as well. In a white paper, I defined embedded finance as "the strategic integration of financial services into nonbanking client journeys. It transcends the conventional boundaries of a bank by offering services directly within platforms, often without redirecting users to traditional banking interfaces. It offers financial products and services to customers and prospects through a partner not working in banking or insurance. It focuses more on the partnership to connect to non-banking client journeys".

Open banking and embedded finance redefine retail and corporate banking. Banks use them to improve customer experience, find new revenue streams, and attract new customers to grow their top line. Embedded finance is seen as a way to sell financial products through (digital) resellers. Corporate bank clients now start offering credits to their clients by simply plugging in their bank's API. The bank handles the rest, from technology and credit scoring to compliance and customer support.

That wisdom comes with age. It reminds me of an interview I took in July 2019, with Karin Van Hoecke, General Manager Digital Transformation at KBC Bank. In that interview, she shared KBC Bank's ultimate mission with open banking and embedded finance (although it wasn't called that way back then): "The more you can create this ultimate laziness for the customer, without the need to really be there, the more they will appreciate you. In creating this kind of experience, you also achieve the new loyalty. That is certainly a big trend" (Coeckelbergs, The Open Banking Interviews:

Karin Van Hoecke, General Manager Digital Transformation, KBC Bank & Insurance Belgium, 2019).

That ultimate laziness for the customers adds more challenges to the bank, and, if not managed well, many more questions for the customer. That is a learning for financial institutions and clients alike.

The future of banking will be branchless or less-branch at best. Invisible banking is what creates the ultimate laziness: be there when the customer needs to, and don't expect them to make an appointment. Branches are expensive to maintain, and many people prefer to use digital channels rather than traditional branches. As a result, banks are increasingly closing branches and moving towards digital banking. However, this shift in banking services has raised questions about the relationship between banks and their customers. How has it impacted their connection, and what does it mean for the future of banking? We conducted a few interviews with banks that are developing embedded finance offerings and also carried out some online surveys. One of the surveys focused on the biggest challenge in offering embedded finance. The majority, 55%, cited compliance as the biggest challenge. This indicates that banks must redefine the rules of the game to ensure compliance remains a priority while also keeping their customers and partners satisfied.

Linking it back to the previous chapter, all this comes with new risks and uncertainties and requires a new skillset for banks. Most banks had APIs somewhere in their architecture but couldn't grasp the full benefits and knowledge because the understanding of these technologies was limited to a few technical people. With the rise of open banking and embedded finance, businesspeople became interested and started looking for compelling use cases. They learned how to use APIs, and understood the potential and pitfalls. It opened doors to completely new services.

Building successful partnerships was the new mantra. Until a few years ago, banks would mostly look at Fintech providers, but with embedded finance, the provision of financial products and services in non-financial customer journeys, banks also look at other industries. It adds new layers to risk management, branding, and customer care. How do you ensure your brand promise in the third-party customer journey? How do you ensure your partners comply with your brand promise, and what does customer care look like in that context?

Realising dreams today happens online, whether that dream is in the physical or virtual world, whether the realisation of that dream happens on a bank platform or not.

Partnerships can be set up in many forms. I often refer to how we as humans set up partnerships (Coeckelbergs, How to find great collaboration between fintech and banks, 2019):

- Acquaintance: there is no real friendship or deep connection. Third-party providers connecting to the PSD2 APIs, simply because they can, is a good example of that, especially if that connection is set up through an aggregator, a middleman that ensures good functioning between the bank and the fintech.
- Friendship: limited negotiations are required, but you must know each other better before connecting. Fintech will directly connect to the bank for higher quality, maybe richer data, or better services. A friendship could be the start of something bigger. An example would be an age verification API that requires a stricter onboarding process than the standard PSD2 APIs.
- Friendship with benefits: these relationships become more complex and can bring more fulfilment and frustration. To avoid the usual disappointment in a friend-with-benefits relationship, strong relationship management is required, especially in dealing with "polyamorous banks", banks that connect with many (fintech) companies and potentially with direct competitors as well. A good example here is the context of embedded finance, where a bank would provide its services in the process of a third party, allowing that party to offer it to its clients, for example, a merchant offering consumer loans in the POS experience.
- The committed relationship/the wedding: this kind of collaboration touched the bank at its heart. Core banking platforms, payment processors, and wealth management platforms are good examples of that, as are Regtech companies and fintech companies that assist the bank in compliance-related matters. They partner with a bank for a multi-year deal to build or provide strategic components of a bank.

The deeper integrations go, the more important it is to set clear guidelines for both the partners and the end-user. With fintech companies or other organisations connecting to banks to improve the financial services of the end client, it becomes more complicated for banks to properly understand who they are actually banking with and who to contact in case things go wrong.

Imagine you are on holiday, and you realise that your card is not activated to make payments in the country you are visiting. Should you contact your bank or the service provider responsible for activating your card?

My bank's app offers public transport and parking features. Who should I contact if there is an issue with these features? If I receive a parking fine due to the app's malfunction, can I ask the bank to pay for it? If I contact the bank for help, but they cannot assist me, what does that mean for my trust in the bank?

One last example: suppose a bank integrated a Buy-Now-Pay-Later (BNPL) provider, and I used it to delay the payment of a purchase, but I'm unable to pay on time because of a technical problem. Who should I contact: the bank? The BNPL provider? Or the credit services management company that took over after the late payment suddenly became a credit?

All these questions add to the many questions people already have today in managing their money.

People have never been closer to their money; they can constantly see their money movements, they can constantly move money around, and they have more tools available to manage their financial situation in the bank environment and beyond. Still, people are getting detached from their money. Because everything happens online and changes incredibly fast, there is a minority risk of dropping out.

That also means that the more banks provide tools to make people more responsible in managing their financial lives, the more responsibility banks have to ensure everyone remains included, both financially and digitally.

Banking is about realising dreams, everybody's dreams...

3.7 Conclusion

From the very beginning of computing power, the invention of magnetic stripes or chips, banks have always been at the forefront of innovation and digitalisation. The many challenges on their plate made them lose some of their positioning over the past 20 years. The democratisation of software development, the creativity of fintech entrepreneurs, and the regulatory context put banks in a defensive position. The focus in digital transformation was too often to create synergies and efficiency, for employees and also customers, whether they asked for it or not.

So after the fall of trust with the GFC, a few scandals, and the consolidation that made banks focus more on scale than on their customers, digital transformation added to the fall of trust in many markets as people didn't see what was in it for them.

Efficiency and convenience were the key reasons for setting up internet banking and, later, mobile banking. These services transformed the face of banking and built on its existing core. They are an evolution, not a revolution.

What it meant is that banking became more rational, and banks created more emotional distance with their customers. That was never the plan; after all, on each mobile phone, banks were never closer to their customers and never more available. Yet, digital banking never got to the level of emotionally connecting to people.

In the past, people trusted their financial advisor, someone with a branch in their local area. These advisors knew their clients and their stories, and there was a strong personal relationship between them. People felt comfortable trusting their money to these individuals, as they were known to be reliable, honest, and fair. However, it's important to note that there were many fraud cases during this time period, as these branch managers were difficult to control.

Despite this, people valued the personal contact they had with their financial advisors. They could speak with a real person rather than relying on screens or apps. Nowadays, people often must put their trust in larger organisations instead of individuals, which can be more difficult. Therefore, it is crucial for banks and financial institutions to demonstrate their commitment to their customers' needs and show that they are reliable, honest, and fair.

As banks have attempted to reduce costs, they have closed many local branches, and the personal connection between customers and financial advisors has become less common. We clearly don't want to go back to a time without digital banking. Still, it is important to understand how this changed the relationship between banks and people, as this is part of the answer in rewriting the narrative in banking. This new narrative will demand banks to find out how to define their ethos.

Some banks are already rewriting their narrative. The belief that the ultimate laziness is the new loyalty is a strong message. It is a commitment to bring digital banking to such a level that by not being there, you connect with customers. This means that as a bank, you are the invisible hand that helps your customers. And it means that this support is so valuable that no other value proposition can convince you to switch.

Open banking and embedded finance are some tools that can help with that. Much more will be needed to shift the industry from efficiency by design to effectiveness by design, but it is a shift required to get everyone on board banking in a digital society. Effectiveness by design means that digital banking is easy to use for everyone. The industry will need to build more complex solutions to remove complexity from the end-user. More importantly, they will need to master that complexity to avoid ending up using it as an excuse for why people have no trust or hope in a digital banking environment.

Notes

1. Nickel is a French neobank launched in 2014 whose growth strategy is based on partnerships with small merchants, often press shops. At the date of writing, they claimed to have a network of 10,000 Nickel point in Europe. 10 years after launch, they have + 3.5 million clients in Europe, mainly in France, Belgium Germany, Portugal and Spain.
2. PSD3 and PSR are a natural evolution of PSD2. The European Commission concluded that PSD2 made payments more secure, less sensitive for fraud and boost for innovation, but more could be done. PSD3 will focus on the lications and operations of payment services providers and needs to be incorporated into local legislation, whereas PSR is a regulation that covers the responsibilities of payment providers in the context of fraud, security, consumer rights, and improvements of the functioning of open banking (European Commission, 2023).
3. FIDA is a new framework in the EU for the responsible sharing of financial data, not just payment data, but much more, including insurance, investments.... It is expected to accelerate innovation in the finance industry further and stimulate competition.

4

Ethics, Ethos, and the Rhetoric Triangle

4.1 Introduction

Banks have undergone an incredible transformation. Their history has demonstrated that deregulation, leaving too much freedom, can lead to greed, a feeling of superiority, and abuse, but also that overregulation can be damaging to the industry.

Banks translated digital customer (and bank) expectations in their products and services, which added to the complexity of what banking has become despite the increased convenience and efficiency it brought. Where banking used to be a people business, it has evolved increasingly into a transactional business. Banking became commoditised. As much as the industry and I say that banking is about realising dreams, the average person on the street will disagree with me. Most people still trust banks with their money, but that doesn't mean they believe in banks.

Something is broken in the banking industry. The narrative of banking is no longer compelling. This chapter aims to provide a framework that helps you understand how the many hints I have given so far are affecting that narrative and what can be done to change it.

Many banks struggle to communicate their strategy in a convincing way and have a hard time explaining that their actions have a higher goal beyond pure profits and shareholder returns. They often fail to convince people of the value they provide, leading to a lack of trust and likability. They fail in their evangelisation that banking is more than simple payments, safeguarding of assets, and loan issuing.

R. Coeckelbergs, *A New Ethos in Banking*,
https://doi.org/10.1007/978-3-031-78234-3_4

As someone who facilitates networking in the banking industry and frequently blogs about its future, I am exposed to jokes and remarks about the industry, which provides me with insights into society's perception of banking. That perception isn't great. Banks need to rethink their narrative to avoid the people feeling fed up "contributing to the banks' profits given the little earnings they get on the savings account".

People read about high profits that are not high compared to other industries, by the way. People read about the high CEO salaries, which are no longer much higher than what other industries would pay. But they read those numbers in their personal context, a context that is heavily influenced by the recent history of banking and negative press attention. These people didn't forget that many banks needed to be saved during the Global Financial Crisis. They didn't forget the money laundering practices and the role of financial institutions in terrorist financing, as they don't consider how much of that is out of the control of most financial institutions. They didn't forget the low returns on their savings.

Agreed. They didn't consider how much of that is due to a monetary strategy defined by central banks. These people read the profits and compare them with what they get at the end of the year, and banks are incapable of explaining to these people that this is normal. That is where often the problem lies; that is partly why a new ethos is needed. Too many examples in recent history have broken trust in financial services, from isolated cases of terrorist financing, fraud, and speculative behaviour to the financial crisis, which required governments to save banks.

People leading these companies have lost their compass and their eyes on reality. They forgot that a solid corporate business is built on purpose and a clear set of rules and ethical guidelines. It happens in every industry, but I believe searching for a new ethos is more pressing in banking than in any other industry because banking affects everything. Banks make the money go round, and that affects which industry thrives, how fast our world evolves, and how healthy our society remains. That is why banking needs a new ethos, more than any other industry.

Some may argue that banks are not the value creators that they often claim to be (Mazzucato, The value of everything—Making and taking in the global economy, 2019, pp. 22–74), and there are very valid points to make that case, I am convinced of the importance of banks in the current construct of society.

A friend once told me that banks define how our society and communities will look in 10 years. Let that sink in for a bit.

Banks define how our society and communities will look in 10 years. Financial institutions approve which project can get additional funding through loans. Banks define which entrepreneur can realise their dream, meaning they also influence employment opportunities and welfare. Banks determine in their credit policy who can get a mortgage and under which conditions, and a bad personal credit strategy may even define how poverty evolves in the worst case. Look at the sustainability transition, where banks today favour people who go the extra mile to reduce the house's carbon footprint they signed a mortgage for. Banks influence the speed of change in that transition.

The sceptic will say that if banks don't do it, venture capitalists, alternative lending providers, or other shadow banks will do the trick, but that, in my opinion, is yet another argument to say why banks need to redefine their ethos. As regulated entities, they are best positioned to realise dreams, in a transparent way with products that are relatively easy to understand and even easier to have access to.

Banks are the veins that let money circulate. Intermediation from short-term excess into long-term needs assists a society in growing and evolving, hopefully into its better self. Turning societies into their better selves requires a moral compass, a business built on purpose, and a clear set of rules and ethical guidelines from everyone involved, including banks.

If we want to define the new ethos in banking, we need to agree on a couple of definitions. I'm not a philosopher by profession. I'm not an ethicist either, so I didn't want to write a book on ethics in banking. There are others who are much better at doing that. But of course, as I'm writing about it, I cannot ignore some of the fundamentals either.

Ethics and ethos are two different concepts from the same bloodline. And because ethos is part of a trinity, we should also look at pathos and logos. What's more, this trinity will form the basis for the rest of the book.

To ensure a common understanding of ethos at our most prominent event in 2023, The Banking Scene Conference Brussels on May 16, we asked the audience what came top of mind when they heard "A New Ethos in Banking". 73 attendees shared 95 answers, and this was the feedback:

The answers were very diverse, and rightfully so. Most of them related to ESG themes like sustainability, responsible banking, technology, AI, and business model change, but also transparency and fair banking. Most answers were pretty much linked to the subtitle of this book, "Embracing Values and Ethics for a Meaningful Transformation".

One word was missing: no one answered ethics. Ethics didn't come top of mind when they heard "A New Ethos in Banking". Is that because they interpreted ethos and ethics as one and the same word? Is it because the survey wasn't done in an English-native country that saw ethos as a "poetic license"?

Or did everyone know that ethos and ethics are two completely different concepts from the same family? Indeed, ethos and ethics are different, which is also why the book may come over as much less theoretical or idealistic than you'd first expect.

In this chapter, I will clarify what the words ethics, ethos, pathos, and logos mean, what they mean to banking, and why they are so important in rewriting the narrative of banking.

4.2 It All Starts with Ethics

The very basics of ethos is ethics. It's a modern word that is regularly used in today's business and ESG context. The word ethics is derived from the Greek "ethiké", meaning morality, showing moral character.

One of the first philosophers in the field of "ethiké" was Socrates, who believed that a person will naturally do good if they know what is right. Evil or bad actions are the result of ignorance. Bankers, for example, who are fully aware of the consequences of their actions whether that be on their situation, their organisation, and society, would not commit any crimes or act against the ethical framework within which they operate.

For a more modern definition, I had a look at The Cambridge Dictionary (The Cambridge Dictionary, 2023): "ethics is the study of what is morally right and wrong, or a set of beliefs about what is morally right and wrong". The number of books and studies defining ethics is endless, and since this is not a book on ethics, I suggest restricting our usage of ethics, for the sake of this book, to the definition as just explained.

Close to 2,500 years and many revisions on ethics after the first Greek philosophers, in the early 1970s, the term business ethics came into common use. In the book "Capitalism and Freedom" or 1962, Nobel Prize winner Milton Friedman explained ethics as "the one and only one social responsibility of business: to use resources and engage in activities designed to increase its profits so long as it stays within the rule of the game, which is to say, engages in open and free competition, without deception or fraud".

Close to ten years earlier, in 1953, Howard R. Bowen wrote the book Social Responsibilities of the Businessman, explaining CSR (Corporate Social Responsibility) as "the obligation of businessmen to pursue those policies, to make those decisions, or to follow those lines of action that are desirable in terms of the objectives and values of our society" (Rovera, 2022).

Business ethics have evolved and matured over the recent years. Milton Friedman believed that a corporate executive's only purpose is to make as much money for its shareholders as possible, confirming the basic rules of society, both those embodied in law and those embodied in ethical custom. He believed that ethics can only exist on an individual basis.

In the 1980s and 1990s, business ethics became a popular practice in the USA, a way for many to distance themselves from the business scandals of the day, what happened during the savings and loan crisis in the United States that led to the failure of 32% of the savings and loan associations in the United States between 1986 to 1995 because of a multitude of reasons, including the quickly changing economic environment, deregulation and the resulting transformation of businesses looking at more lucrative and riskier products, bad management of bank's credit risks (Federal Deposit Insurance Corporation, 2023).

Over the years, this concept evolved, and every reputable organisation adopted business ethics as an integral part of establishing governance. Many organisations have a code of conduct that reflects their Corporate Social Responsibility, a term that was also embraced, stimulated, and explained by, for example, the European Commission (2001) as "a concept whereby companies integrate social and environmental concerns in their business operations and their interaction with their stakeholders on a voluntary basis. Corporate social responsibility concerns actions by companies over and above their legal obligations towards society and the environment" (European Commission, 2011).

Companies write down their CSR strategy to ensure it is tangible and that there is a common understanding of how the organisation looks at what is good and wrong. Topics that are important in society and values that are commonly embraced are high on the CSR agenda. That means that today, from a Corporate Social Responsibility point of view, we see an increased focus on sustainability. Whether they apply the "tick the box principle" or they are effectively adding it to their corporate DNA, organisations put much more attention than in the (far) past.

Today, it is a common good to think beyond shareholder value. Organisations look at a deeper purpose, and many international organisations lead the way in leaving a positive footprint for people, society, and the planet, in some regions more than in others.

The United Nations, for example, defined 17 Social Development Goals (SDGs) to be implemented from 2016 to 2030 as the biggest joint objective of the international community. Climate action is one such objective. Ending poverty and hunger are two other objectives. These objectives are a common compass for organisations that write down their code of conduct. It is seen as the right thing to do.

ESG is another acronym that gained a lot of attention lately. It stands for Environmental, Social, and Corporate Governance. It covers a broad spectrum of attention points to ensure an organisation is not just making profits but also contributing to a better society and the planet at large.

Specifically in banking, it is worth mentioning Christina Rovera's interpretation of John Stuart Mill on what is good and wrong for banking, who said Mill measures right and wrong action according to the moral or ethical quality of the action itself. "A bank is therefore moral when it promotes the happiness of many. This includes lending to companies that "do not produce the opposite of happiness", i.e. that meet the ESG criteria" (Cristina, 2022).

Sustainability became an essential pillar in defining the ethical framework the financial world must adhere to. ESG is the broader spectrum that combines Environmental, Social Development, and Governance measures to define an organisation's business ethics. ESG also forms the basis for Sustainable and responsible investments (SRI). The lack of transparency led to the publication of the Sustainable Finance Disclose in Europe (SFDR). More on that later.

Similarly, the fight against terrorist financing and fraud is the result of the understanding that organised crime is bad and can only survive with access to financial services. The IMF transparently shares on its website that this issue got much more attention after the attacks on the World Trade Center: "The IMF expanded its AML efforts in 2000 and extended them to CFT after the terrorist attacks on September 11, 2001. In 2004, the IMF Executive Board agreed to make AML/CFT assessments and capacity development a regular part of IMF work. In 2018, as part of its five-year review cycle of policy, the IMF Executive Board reviewed the IMF's AML/CFT strategy and gave strategic directions for the work ahead" (International Monetary Fund, 2023). This shows that ethics is a living science and will be interpreted in the context of place and time, meaning that this framework may look different from culture to culture, from religion to religion, from country to country, and from decade to decade.

Despite all this, the banking sector remains an industry of distrust. The many regulations and interventions have provided only little help so far for the banking industry to recover from past scandals. The paper "Virtuous Banking—Placing Ethos and Purpose at the Heart of Finance" described this nicely: "The Banking sector had lost its ethos. Instead of a culture that prioritises its own self-interest, it should look towards the fulfilment of a broader common good and its wider social purpose" (Llewellyn et al., 2014).

The paper also states that "most ethical systems are formal and empty, and they assume that merely following the rules will ensure a good outcome. But virtue is about content and the ability to understand the spirit, not the letter of the law. Given the failure of rule-following ethics as evidenced by the crash and multiple scandals, banks evidently, and most obviously, need to be of good character. Bankers need to know the good, do the good and be good".

To summarise, banks lost their ethos and, in a way, also ignored the ethical framework that defined their rules of a fair game. To understand the fundamental problem, it is important to understand the difference between ethics and ethos.

Now we have a clearer idea of ethics, let's have a look at Ethos.

4.3 The Soul: Ethos

The basic understanding of what is good and wrong are the fundamental building blocks for everything, but it is not enough. Ethics is a more scientific approach, a study. It allows us to balance arguments against each other in ethical dilemmas, but often we'll need more to choose sides in such a dilemma. We need more than an understanding of good and wrong. We need a soul to help us weigh the arguments.

Regarding the good and wrong, what matters is how we live it, how we breathe it, and people around us feel how we live it. That matters because that is what people remember. It is no coincidence that Aristotle came up with "The Rhetoric Triangle", also called the model of persuasion. The three corners of that triangle are also the three elements of the model of persuasion: ethos, pathos, and logos. It is commonly used as a framework for constructive persuasive communication. It can also be a helpful framework to explain how to rewrite a narrative that people accept, one that makes them feel understood and perhaps even involved.

That involvement, or lack thereof, is what the industry is currently struggling with. That is why the rest of this chapter, and the rest of the book, will circle around those three elements and how to get them in balance.

The first element of the triangle, the one I would put on top, is "ethos".

Both ethics and ethos have the same Greek foundation. The Greek word ethos means "custom, character, habit", and ethos (from which ethics is derived) is an adjective derived from the noun ethos (Heimbach, 2015).

The Cambridge Dictionary defines ethos today as "a set of beliefs, ideas, etc., about the social behaviour and relationship of a person or group" (Cambridge Dictionary, 2023), which is commonly known as morality. Often used in rhetoric, it refers to a mode of persuasion, discussed in Aristotle's "Rhetoric", as a component of argument and involves moral competence, expertise, and knowledge. Aristotle explained that the intelligence and competence of a speaker, but also the way that the speaker can share its arguments in a clear way are fundamental, just like the goodwill, the genuine kindness, positive regard, and helpful intentions (Rapp, 2023).

Ethics lays the foundation for a society on what is right and wrong, and ethos is the colour that people and organisations like to give to right and wrong in their behaviour and how they translate that into their messaging and in their actions. A bank can say that it takes anti-money laundering seriously because it is what they are expected to say, but it will lose its credibility if it is fined millions for not taking the necessary measures to avoid money laundering. The message gets lost, and the bank loses its ethos.

That is why I like to refer to the soul of an organisation in the context of ethos.

We know what is good and wrong; society puts rules in place to defend those ethical principles; principles, by the way, that are different from context to context. Just look at, for example, Islamic Finance, which has a different ethical compass compared to the kind of banking most of us consider normal.[1] Over time, we see those societal dynamics in the '80s were completely different from what we feel is right or wrong today, with the emphasis on sustainability as a striking example today, or the evolution from shareholder value creation to broader stakeholder value creation that we saw in recent years.

What's interesting is that an organisation cannot control ethos. As the organisation's soul, it expresses itself through connection with others. It belongs to the audience; ethos is about how the audience looks at you. Take the example of a bully, a not-so-kind person. As much as he or she may believe what he or she does is the right thing, others will determine whether he or she is a bully or not.

That means that ethos in banking does not belong to the bank but to the clients and stakeholders, appealing to what banks bring and contribute. Clients and stakeholders will evaluate how well you convince them to be a trusted partner with the correct ethical values.

Banks can influence their ethos but cannot define it. That is perhaps why Michael Anseeuw, CEO of BNP Paribas Fortis, shared in his interview on ethos in banking that two dimensions need to be in sync to work: an internal and an external dimension. The internal dimension translates into a company's values and culture. In contrast, the external dimension is how you interact with and position yourself towards your customers and society as a bank and an organisation.

He said: "A company where they (the internal and external dimensions) are not in sync will never fly. Then you get in a situation where the values an organisation defines internally are perceived as fake by the outside world. To maintain your integrity and earn people's trust, it's important to have them in sync. In the financial world, we have a specific challenge: balancing our rational and economic reasoning with our role in society. What we do outside should match with how we act inside our organisations and the other way around. And that's not always easy. We cannot limit ourselves to a pure economic story. We must also write a societal story. This means that banks must look differently at doing business than other industries" (Coeckelbergs, About Mary Poppins and the New Ethos in Banking, 2023).

Business Philosopher Jochanan Eynikel explains these internal and external dimensions, the undercurrent and mainstream, in organisations. The undercurrent refers to the organisation's mission, values, image, inspiration, and motivation, and the mainstream refers to the organisation's products and services, communication, and strategy and the infrastructure that supports it (Eynikel, Check-in: The search for meaning and purpose in business, 2021).

This explains the tremendous challenges in defining a new ethos in banking because first, you need to shift the internal alignment and truly embed sometimes new values in the organisation, which could be equally challenging than to convince the outside world that what you did internally is also an authentic and reliable new direction.

A common thread in my banking conversations is that ethos is culture; others refer to "character". Both answers are more similar than you may think. It means that ethos is much larger than ethics. It is not just a set of guiding principles. No, it is part of an organisation's identity. Ethos identifies who that organisation is, who its people are, and which values they represent. Good culture or bad culture is not something that can be described as a general truth. It is a translation of how others, the outside world, believe you behave. No organisation will call itself, for example, having a toxic culture; that is what the outside world makes of it, based on stories, gossip, perception, and testimonials. The other way around, as well.

A strong ethos will make a bank much more resilient as it considers more stakeholders involved in its day-to-day decision-making. Having more trust from these stakeholders will make it much easier to get the right thing done without spending energy on distractions like explaining over and over again why it would be the right thing to do. A strong ethos will align employee visions in the same direction, which in return brings more self-organisation, which yet again adds to a bank's resiliency.

A strong ethos starts with a common understanding of the importance of doing good and acting on that belief. The people in the organisation feel a sense of meaning and are aligned with that meaning. The meaning will create an intrinsic motivation to apply the organisation's values and culture. To define "meaning", Eynikel defined meaning as a matter of experiencing a connection between ourselves, others, and the world. He acknowledged that personal experience is crucial; it is about experiencing that something forms part of a meaningful whole. Companies wishing to embrace ethics and values for a meaningful transformation must establish meaningful connections between their values, products, customers' expectations, and society's needs (Eynikel, The meaning of meaning, 2021).

Klaas Ariaans joined us one day at a round table; he was still Managing Director of Consumer Clients at ABN ARM Bank to explain his self-organisation experiment. He defined it as follows: "within a framework, people can decide how to organise their work to end up with a more efficient and effective organisation, in combination with a better work environment for both employees and customers". In his complaints department, for example, his team decided to give employees the mandate to compensate or solve complaints straight away, up to 500 euros. Although it led to an increase in the pay-outs, it also contributed to more transparency in what was going wrong, and people came up with great ideas to solve it to improve the budgets in the longer term. It is the creativity that will keep an organisation going instead of freezing in case of unexpected problems, and it is a good example of how a strong ethos also improves resilience. None of that additional ownership to employees would be possible without a clear moral compass within the organisation (Coeckelbergs, ABN AMRO Bank's sales transformation with self-organising teams, 2022).

In that respect, the mission to embrace ethics and values is not a marketing gimmick to show the world you are doing good. Ultimately it is landmark of what you as an organisation stand for, and it has a positive economic and financial impact on the organisation. Studies have shown that firms where strong levels in the purpose of the organisation are being evangelised by midlevel employees experience a better performance (Gartenberg et al., 2019).

That is why this debate is much more than a debate on values and doing good. It is also a strategic business question in today's world that affects both how people look at a bank and how a bank can build a solid business in that context. It is much more about the building than it is about explaining the building process. Banks lacking ethos often miss the trust required to function properly as financial institutions. They may not attract the deposits to issue loans or, even worse, lose access to the interbank market, leading to a liquidity shortage and preventing them from fulfilling their obligations to other banks and clients.

A good corporate understanding of its identity, values, and purpose for being is not just a marketing story but an economic reality. Having these elements clear will help organisations connect with their stakeholders much easier, as there is a common ground for dialogue and doing business. Others may also have more goodwill when things to go as expected.

Silicon Valley Bank is a recent example of a bank lacking a strong ethos. One very influential person was enough to kick off an online bank run simply by stating on social media what was at stake, leading to the fall of the banks a short while later. Because banks are so interlinked, a justified rumour (or unjustified, but very convincing, for that matter) about a bank is enough for others to withdraw their funding, leaving the bank behind and completely illiquid. The moment a bank no longer gets the trust of its clients or the banking industry at large, it is doomed to fail.

When we asked Niccolo Polli, Head of Sustainability and Strategy, HSBC Europe, what banks should be looking for in their quest for a new ethos in banking, he said: "To me, this refers to a cultural change that is required in the industry. In that context, the key question is: what is wrong with the current culture in the banking industry, and how did we get here?".

Indeed, the banking history doesn't play in their favour, unfortunately. I speak to banks daily, and I see and feel that they look different from their metier than in the past, but we all know that trust is left by horse and comes back on foot. That is why, too often, the press focuses on the financial results instead of societal contributions and prefers to highlight yet another fraud accusation instead of a bank's support of microlending and its impact on the economy, as many banks do.

That bad press is also ethos; it is a very hard narrative to change, but one that must change to adapt to the attitude of the Average Joe towards banks.

Corin Millais, Head of Socially Responsible Banking at Teachers Mutual Bank, shared his personal reflections on "ethos in banking". He reasoned that "purpose-washing" is part of why the press doesn't talk about it much. Because of the past, people don't believe in a bank's present purpose.

"Culture defines behaviour and actions, and we've seen what culture does", he said. "It's been a global phenomenon of unparalleled greed and growth with a byzantine world of complex structured securities, sub-prime, and in the end, assets and liabilities that simply could not be traded, and that, of course, was the GFC (Global Financial Crisis). Banking was so far away from people and communities that we all paid the price" (Coeckelbergs, About David, Goliath and the new ethos in banking, 2023).

So again: ethos belongs to the audience.

Banks lost connection with their audience because of a wrong culture that stimulated individual success and shareholder value; it was never enough. When the market was saturated, they developed new, more sophisticated products and created new markets to continue and even speed up growth. Because of that, their audience also lost connection with banks.

That explains why we cannot talk about ethos without digging deeper into logos and pathos.

4.4 The Mind: Logos

Aristotle was a wise man who understood the "rule of three", a powerful cognitive concept that harnesses the cognitive strengths of pattern recognition. He invented the rhetoric triangle, a model for effective communication, which he believed to be a combination of three elements that need to be in balance. The first one is "Ethos. The second element in the rhetoric triangle is "Logos". Logos refers to reasoning based on scientific numbers and rational thinking.

The word comes from Ancient Greek and was reused in Latin, where it became "Legus", referring to "to put in order, arrange, gather count", and in modern usages, it connotes the verbs "account", "reason", or "discourse"(Liddell & Scott, 1889). Adding rational elements to a discourse can make reasoning truly powerful as long as the reasoning is digestible by the audience, readers, or listeners. When the reasoning gets too complex, the storyteller will lose connection with the audience.

Banking is all about financials and numbers. Banks have always been great at it. They were among the first to adopt double accounting to control their business. They take on short-term deposits and transform them into long-term loans, closely monitoring the risks involved to ensure everyone benefits.

Over the past decades, that logic has evolved to the extreme, in a perfect storm of elements that all make sense in their own context, but became a toxic cocktail of hallucinations.

Products became extremely complex and lacked transparency and a clear understanding of their construction. A fear of missing out, more so than the prospect of a clear and beneficial business case, was the main driver for many banks to engage in these new products.

On top of that, banks kept growing because, according to the theory, scale equals lower cost equals higher profitability. So, banks broadened their business and needed to learn new skills to become universal banks. They got bigger, often too big, and they became more complex, with too many business lines and a lack of overview and internal oversight. Acquisitions in new geographies increased legal, fiscal, and regulatory complexity, combined with the challenge of aligning multiple cultures into one, which may be the hardest challenge.

Economies of scale and acquisitions were discussed and approved in board meetings full of smart people, kindly assisted and guided by equally smart management consultants. In exactly that vacuum, rational thinking is mostly appreciated, but guess what I learned in a speech on nudging recently? In these kinds of meetings, rational thinking is highly overrated, and emotional evaluations are mostly underrated. So, you come to a contradictive situation where people assume everything is based on rational theories. They read in books or learned at business schools, but they made their decisions based mostly on gut feeling and were biased by others.

Merging two institutions into one is not a simple 1 + 1 = 2, and the biggest challenge is human alignment, making sure everyone looks in the same direction. It is emotionally driven. So, banks thought they respected logos in their management but lost track without knowing it. Realising economies of scale by bringing all group entities into one organisation or on one platform can be expensive and challenging, and sometimes even impossible because of one key struggle: aligning the people to do the integrations and to adhere to the common processes, procedures, and internal culture.

The banking industry's network of banks and their dependence on each other made it unique from all other industries. By adding 0s and 1s to the equation, all these assumed rational reasonings led to an explosive cocktail, showing that perhaps the industry went too far in its rationalisation process, pushing logos to the extreme.

As financial services grew in importance, their health determined society's health. The GFC demonstrated our dependence on the financial industry, and that is a scary thought because its complexity is intangible for most people. That is one of the reasons why banks lost connection with society. Products got overly complex, leaving people to choose things they didn't understand with the promise of making a good expected profit, which was understood as "the promise of good profit". This all went well until it didn't. Despite all the rational logic to explain what went wrong, banks couldn't connect with their clients anymore, nor with most parts of society either.

Banks got lost in their own complexity, in their own rationalisation, in their own reality.

That is where regulators stepped up and enforced measures like MiFID and MiFID2 to protect customers. With the best intentions, they pointed banks to their duty of care. Regulations were set up to ensure that banks understand people's expectations of their finances and properly explain the "logos" behind a product. On top of that, the new regulations also expect to get rational measurements to ensure the bank "feels" with its customers. MiFID2 obliges, for example, European banks to regularly survey their clients on their financial knowledge and risk appetite. Since August 2, 2022, sustainability preferences have also been included. It also creates much more transparency in the pricing structure banks should apply to their customers.

It is a very rational approach to align banks and their clients, but it isn't always reaching its desired outcome. There is a reason why organisations like the European Banking Federation are putting so much emphasis on financial education, like many other financial sector federations across the world. Over the years, the complexity of banking and financial services didn't translate into more financial awareness and training at school.

According to Aristotle, logos, or logical reasoning, is not as effective without a common belief that the speaker, or counterparty, can be trusted and is credible, known as ethos. To translate that rhetoric theory into banking practice, banks cannot solely rely on logical arguments to explain their importance and how well they serve people, especially not after the history of mismanagement of risks and the delusions of grandeur in the past. If the logical arguments are not a logical extension of the characteristic of the organisation or person behind it, the arguments will be rejected.

Some bankers believe they must reframe their story to solve this issue, but that is only part of the solution. Banks must not only say they are trustworthy but also demonstrate it consistently, over and over again. Actions and words must be aligned, or like Michael Anseeuw, CEO of BNP Paribas Fortis told me: "The internal dimension translates into a company's values and culture. In contrast, the external dimension is how you interact with and position yourself towards your customers and society as a bank and an organisation. A company where they (in internal and external dimensions) are not in sync will never fly" (Coeckelbergs, About Mary Poppins and the New Ethos in Banking, 2023).

Having explained Ethos and Logos, we are left with Pathos, the third element.

4.5 The Heart: Pathos

Reasoning and trust are not enough to create an exciting story. As we mentioned, many perceived rational decisions are made with a touch of the heart. That brings us to the third corner of the triangle. A story without pathos is like a cake without sugar: the egg white makes it fluffy, and the flour and milk give it the right texture and satiety, but it will not be appealing to eat.

Pathos "appeals to the emotions and ideas of the audience and elicits a feeling that already resides in them" (Mt Hood Community College, 2019). Like ethos and logos, it comes from the old Greek rhetoric. The emotional appeal makes an audience feel with the orator; it makes a connection.

Deploying pathos means that you look for hooks and ties with your audience that make them feel connected to your story on a deeper personal or emotional level.

In a corporate context, pathos can be interpreted in many ways. Specifically, advertisements are seen as targeting the emotional side to persuade consumers to buy goods and services. In that respect, emotional awareness is critical (Mt Hood Community College, 2019). What are your clients and other stakeholders looking for, and what do they expect from you?

If banking is about realising dreams, then pathos is the capability to understand what those dreams are and how to realise them. In today's digital banking context, organisations need to go beyond building convenience and seamless experiences; they need to make a connection and build loyalty.

In the past, every village and community had its branches, and people knew their personal advisors very personally. That advisor understood his customers' lives, their families, the business they were in, the risks of providing credit, and a lot more. There was a personal, emotional connection. They knew each individual and respected all in their uniqueness. They understood that everyone had their own journey, dreams, and issues.

Often, people choose their bank because it has the closest branch, or because they, or their parents, had the best connection with a specific bank advisor, connection as a unique selling proposition. Banks could afford to offer the same services and products as their competitors because those services didn't sell, it was the connection with an advisor and the distance from a branch that was the USP.

With the rise of the internet, this slowly changed. Banks created a digital environment for day-to-day banking services. With digital banking, that trust element shifted. People could no longer trust their advisor, who disappeared, either because the branch disappeared or because they moved to another branch after yet another reorganisation.

In the digital world, these banks often still offer essentially the same services and products as their peers, but distance is no longer a USP, as every app is one click away. What's more, digital technology has not really transformed banking so far; it merely digitised existing processes but didn't fundamentally change a bank's role.

People used to go to a branch to get cash in a branch, they used to make a credit transfer in a bank, they used to invest their money in a branch. Today, they do this in digital channels; in the best case, it doesn't lead to unnecessary paperwork at the end of the process.

That switch had serious consequences for pathos in the rhetoric triangle. The longer I think about it, the more I'm convinced that the industry underestimated the impact on its customers. We should reconsider how we think about trust. At The Banking Scene, I had a year where trust was the key theme for my events. Bankers joined on stage, explaining that some of the trust was gone because of past misbehaviour by some institutions and that they need to ensure that they build trust again.

As I look at this right now, I believe there are two mistakes in that sentence. The first one is that banks cannot rebuild trust; it is an external evolution that banks can merely influence but not create themselves. The other one is that trust was gone because of past misbehaviour by some institutions, and I'm sure it affected the industry: history has shown that it is something of all times. However, banks were not alone in having serious crises that affected trust; they seemed to be the ones that suffered the longest from mistrust after a crisis.

Perhaps the root cause is a combination of past mistakes and the evolution of who to trust. Because of the industry's digital transformation, people have to trust an institution instead of a person. People trust a brand, not a team in the local branch. Banking is no longer a personal business; it is a digital business.

As we used to trust taxi drivers, we trust Uber today. If the taxi driver does something wrong, we blame the organisation behind the platform. We will not reconsider the driver but will reconsider the service provider.

So, if we move from trusting one bank advisor most of our lives to trusting an institution that is pretty much reliant on its leadership, what does that do to a person? How does it affect the understanding of trust? A lot.

Banks often say that they remain a people business for exactly that reason. Trust is a result of bonding, and recreating that bonding in a digital world is a challenge.

Don't get me wrong: there is no going back from this digital transformation. The only way forward is digital, and luckily, I see technology reaching a level where we can finally start thinking of a new phase in digital transformation.

Banks often forget that customer segments are, in the end, a collection of individuals with much more coloured expectations than the descriptions of each segment prescribe. Having worked in the banking industry, I know how superficial these segments sometimes were. All the clients would be divided into categories based on their incomes or Assets Under Management (AUM). But what does that even mean? Many banks lost pathos with their clients, often in combination with, or because of, rational thinking based on wrong assumptions.

No logos, no pathos, no ethos.

Personal banking became impersonal, and the contact between banks and clients often disappeared.

In some countries, this transition from branches to online worked better than in others. But the fewer branches there are, the bigger the rebellion gets against banks. The Netherlands is a country where it worked relatively well. They found ways to balance the branch density and demand to a certain level. What made the Netherlands so unique is that the country started researching digital inclusion very early, as pioneers and set up policy measures from very early on to ensure people were digitally included and equipped with the right level of skills to interact in a digital society (Coeckelbergs, Digital inclusion in Belgium Anno 2022, 2022). Because of that connection with the digital world, it was much easier for banks to roll out their digital strategies, as there was pathos involved: a common understanding of the advantages of digital.

But even in the Netherlands, the banking industry is taking extra measures to ensure everyone keeps having access to banking, even when they are not online. Let's be clear: these are not only the elderly that are being addressed, as Cecile Botden, Senior Customer Journey Expert at ING Netherlands, explained in one of our virtual sessions on financial inclusion: "The elderly are the sub-category that is closest to us. They are top of mind when we think about digital inclusion, but we are not representative of the whole society. That is why it's so important that we listen to a diverse group of customers to understand their needs and build solutions to include them all" (Coeckelbergs, Digital inclusion—Making digital banking available for all, 2022).

Digital banking makes people's lives a lot simpler, rationally speaking. Many studies show that it is an incredible time saver and that it can structure administration for clients to help them improve their financial situation. Unfortunately, this often ignores the emotional aspects of people's reluctance to change. It ignored the individual context of every person using it.

Digital was long seen as a rational exercise, a one-fit-for-all, and underestimated the bigger fear of society towards a digital world. Although that fear is not universal in the banking industry, it simply adds to the many other arguments that confirm the perception that banks don't understand their customers or at least don't want to listen to them from the moment it negatively affects their P&L.

Luckily, today, the industry is taking initiatives to improve financial access. Some banks, like KBC Bank in Belgium, have launched a "KBC Belmobiel" to assist people who struggle with digital banking (KBC Bank, 2024), much like ABN AMRO's financial care coaches in the Netherlands. The bank has a team of certified experts to become more digitally literate, with more than 100, in August 2023 (ABN AMRO Bank N.V., 2023).

Unfortunately, these remain exceptions to the rule.

4.6 The Rhetoric Triangle and the Biological Forms of Trust

At a marketing event, I was honoured to attend a session by Erik Schoppen, a Professor and Lecturer on Brand Management. One of the things he explained was how our brain defines what we trust or not. He shared three biological forms of trust that define whether someone is willing to open themselves to another or not.

The parallels with the rhetoric triangle are striking, and they show that there is a biological explanation behind what I'm explaining in this book: three forms of trust and three elements to master to be trusted.

Cognitive trust, also known as reasoned trust, is the primary biological form of trust. This type of trust is highly fact-based, semantic, calculated, and explicit, with competence being a key factor. Experts rely heavily on this type of trust to convey their message. In building a brand reputation, effective communication is vital, and this type of trust plays a significant role. A brand's reliability, competence, and capabilities are critical factors that need to be established, along with its history, which proves its competence. For instance, banks need to communicate their reputation regarding security, resilience, and performance, as well as ethical conduct and a consistent track record of reliability, to establish cognitive trust. In Aristotle's rhetoric triangle, this element falls under the category of logos.

The second type of trust is called "conative" or "behavioural" trust. Cognitive trust is based on the message, while conative trust is about how the message is conveyed. This type of trust is all about predictability, and it depends on procedural systems, integrity, values, habits, stability, and experience. To build conative trust, it is important to be consistent in one's actions and to fulfil promises. It is also important to ask ourselves whether what we are experiencing is what we expected and whether we are open to accepting it.

Business transformation, whether it involves digitalisation or other changes, can be a challenging task. This is because transformation means change and change means breaking away from predictability. However, if customers understand that quality services will not be affected, they may become more open to change. If you do this well, it can lead to greater loyalty and satisfaction. Those who champion behavioural trust need to worry less about their message, as their clients will spread positive word-of-mouth based on their positive experiences. This form is key to people's user experience. That is also what ethos is about.

Affective trust or emotional trust is the final type of trust. It is more implicit, uncontrolled, and unpredictable. Emotional trust is based on the feelings of connection and confidence that people develop with brands or organisations. It is more about personal experiences, and empathy is the key to maintaining this type of trust. If a brand is unable to connect with its customers on an emotional level, it will be challenging to establish trust. Social interactions are crucial in building emotional trust, as people feel heard and accepted. The keyword for emotional trust is "warmth", as it is essential in building deep and lasting relationships that go beyond the transactional and monetary aspects. Just like a good user experience can help people transition to new digital developments, a strong emotional trust can make customers curious and excited to try new things. This is where pathos comes into play.

He explained that these three forms of trust define someone's willingness to be vulnerable to the actions of another. Once these are fulfilled, trust can exist, which he explained as "the relationship with another identifiable party who is perceived to act and react with volition toward the trustor. This is the leap of faith" (Schoppen, 2017).

What are the best practices to ensure that people will trust banks again, not just with their money but also intrinsically? What is necessary to create that leap of faith, so people are open to accepting that banks are valuable actors in society? Is it possible for banks to rebuild relationships based on integrity and values so that when banks make a statement, it is not simply ignored or interpreted as greenwashing? Can banks rewrite their narrative in banking to regain the trust of their customers and society?

4.7 Conclusion

This chapter sets a framework for banks and other industries alike to understand what is important in successfully rewriting their narrative. Although we focused on banking, the three keywords we presented can be applied to any industry or situation. These keywords have their origins in ancient Greece, so we are not reinventing the wheel.

Ethos refers to how an audience perceives a spokesperson or organisation, which is determined by their behaviour and attitude. A speaker's credibility, trustworthiness, and expertise are all a part of their ethos. It is important to remember that ethos belongs to the audience. It is their perception of the speaker that ultimately determines their credibility.

Logos refers to a message's rationale. This includes the arguments and statistics used to support the message, as well as the speaker's expertise. The strength of these elements determines how convincing the message is. Logos is essential for a persuasive argument because it provides the logical backbone upon which the argument is based.

Pathos refers to emotional appeal or empathy with the audience. Understanding your audience and their sensitivities is key to creating a successful message. Empathy allows you to connect with your audience on a deeper level, helping you understand their needs, desires, and motivations. A good understanding of your audience's emotional responses can help you craft a message that resonates with them.

The most effective communicators balance all three modes of persuasion. Ethos establishes credibility, logos provides logical support for the message, and pathos connects with the audience on an emotional level. If you can master these three modes of persuasion, you will be able to communicate your message more effectively and persuade your audience to take action. It's important to note that the discussed theory is not just another fluffy framework. By drawing a parallel with biological forms of trust, it becomes clear how effective it is. After reading this chapter, you'll start seeing corporate communication, marketing campaigns, and storytelling differently. You'll also notice how consistent they are with the day-to-day work of the respective organisation.

Soon, you'll realise that some stories feel much more reliable than others. A good understanding of ethos, pathos, and logos will help you understand why some organisations are more successful in sharing their story than others, even when the story is the same.

Effective communication requires more than words. It involves a deep understanding of your audience and the ability to connect with them on different levels. It also involves providing the right answer, both in words and in action, to the desires, expectations, or good understanding of that audience. This is where the three modes of persuasion—ethos, logos, and pathos—come into play.

How to manage your company and guide it through uncharted waters? We have witnessed the digital transformation that has changed how we bank and interact with our bank. However, too often, digital transformation was only focused on efficiency and convenience, and banks forgot the values they represent. This led to a mismatch and imbalance of the three elements—ethos, logos, and pathos.

Banks have developed online banking services and linked them to proximity, claiming they have never been closer to their customers. However, when something goes wrong, they often fail to have a fallback process in place. Contact centres' KPIs typically measure the number of calls per person rather than customer satisfaction. As a result, clients often find it difficult to reach their bank when they need help. If they do, they often lack the quality support they hoped for.

Banks often boast about their contribution to a more sustainable world. However, they typically only offer the same investment and lending products as their competitors, and they don't always practice what they preach.

Finding consistency between internal management and external communication, between what the bank wants and confirming the wishes of the customer and society at large, is tough. It is a frequently misunderstood issue that is still very hard and time-consuming to correct. People still point to the Global Financial Crisis to explain banks cannot be trusted, even though that is close to 20 years ago now.

It takes a constant and consistent effort to make sure these three concepts pathos—ethos—logos are in balance.

Note

1. It is an interest-free, asset-based, and ethically driven banking sytem. They will limit themselves to sharia-compliant services, thus avoiding industries like alcoholic beverages and gambling. Islamic finance has an alternative business model because interest fees are prohibited. They don't issue loans based on the creditworthiness of a customer. Wajeeha Hussain Awadh, Group Chief Digital Officer et al. Baraka Group once explained me: "If I want to deal with a certain customer, it is important for me to know that the business that I am financing is feasible and viable. It is a partnership between the customer and my bank" (Coeckelbergs, Islamic banking lessons and the future of Islamic Fintech, 2020).

5

The Complexity of Putting Ethos in Practice, A Few Use Cases

5.1 Introduction

Understanding ethos, pathos, and logos as individual concepts is easy, but it can be challenging to apply them effectively in practice. The way they are interconnected largely depends on the context, and it is important for every organisation to assess how they can be adapted to their specific needs. It is natural for everyone to want to create a persuasive and appealing narrative for their organisation. However, effectively writing such a narrative is much harder, and not everyone has their faith in their own hands.

This chapter presents various examples to demonstrate how establishing and losing ethos can be apparent in some cases while being abstract in others. It is not a straightforward exercise, and due to its connection with pathos, which deals with emotions, it can be subjective and even emotional.

Because of its nature, the banking industry may be facing even more challenges than any other industry, as they are a money business and people do not always have a very positive relationship with money.

Danske Bank's case is considered to be one of the biggest anti-money laundering cases in Europe. It originated from a small branch in Estonia. This particular case has been marked as one of the largest fraud cases in recent banking history, leading to a complete transformation of the bank over the past decade. In an effort to rewrite its narrative and redefine its ethos, Danske Bank underwent significant changes over the years.

As a second example, let's discuss how the Greek Credit Crisis, which took place right after the Global Financial Crisis, continues to affect the way banks function and promote themselves today. The crisis had a severe effect on the

R. Coeckelbergs, *A New Ethos in Banking*,
https://doi.org/10.1007/978-3-031-78234-3_5

Greek economy and the local banks, and its effects still shape the role of banks in the economy today.

The third instance pertains to the FX mortgages in Poland that were advertised as a pioneering product that guaranteed a bright financial future for many Polish customers at reasonable rates. However, it resulted in a monetary catastrophe caused by a combination of factors, leading to adverse media attention in both local and international news that persists even after numerous years.

The Silicon Valley Bank, a secure place for young start-up and scale-up companies with a particular emphasis on technology, faced a perfect storm in early 2023. Unfortunately, they didn't have the time to rewrite their narrative. This is an intriguing case as their bankruptcy also temporarily affected the ethos of other banks.

5.2 The Danish Connection

Danske Bank is Denmark's biggest bank. Founded in 1871, this institution has been a Danish national symbol for more than 150 years. Today, the bank has activities in 8 countries and is supported by more than 20,000 employees at the time of writing.

5.2.1 The Rise and Fall of a National Monument

Prior to 1871, only two banks in Denmark had the financial capacity to support the agriculture industry. The rest was too small to structurally support agriculture. Most banks were only located in Copenhagen as well. The two banks with sufficient scale were not interested in investing in farming. The establishment of Den Danske Landmandsbank marked a significant shift in Denmark's agricultural landscape. The bank's financial support enabled farmers to adopt modern agricultural practices and expand their businesses through the formation of cooperatives. This led to a rapid evolution of the agriculture industry, resulting in growing economies of scale.

The bank established a reputation as a trustworthy brand for a particular group of customers. Over time, it grew, catering to more people and businesses, and by 1910, it had become one of the largest banks in the Nordics. The bank supported farmers and saw the industry evolve. Farms became bigger and more capital-intensive.

In 1976, the bank changed its name to Den Danske Bank to better reflect its anchoring in Danish society (Danske Bank, 2024).

Despite that position of being a role model for the Danish economy for so many years, the bank wrote one of the darkest pages in the history of European banks as well, ignoring ethical principles for profit. In 2017, it was revealed that Danske's Estonian branch laundered up to €200 billion ($230 billion) of suspicious transactions between 2007 and 2015. The bank advised clients from Russia, Azerbaijan, Georgia, Switzerland, the United States, and Iran on how to avoid an audit trail of bank transfers, how to prepare documents, and set up fiscal structures to launder money. It is said to be the biggest money laundering case in Europe and perhaps even of all time.

The bank settled for $2 billion in the United States for "illegally allowing foreign actors to funnel money through their branch in Estonia to gain unlawful access to the US financial system". In addition, the SEC announced a separate settlement over allegations of money laundering for $413 million (Rabinowitz, 2022).

In Estonia, six employees were charged with laundering $1.6 billion. Prosecutors seized 10 million euros worth of property, which they believed had been bought with the earnings of money laundering services (Reuters, 2023).

Clearly, the money laundering procedures were not respected. There was a disconnect between the Head Office in Denmark and the branch in Tallinn. The physical distance should not have been a problem in today's digital age, but the cultural distance between both was present. The main languages spoken in Tallinn were Russian and Estonian, which made it difficult for the head office, where Danish and English were spoken, to understand the situation fully. This language barrier played a role in the head office's failure to take appropriate action in the first few years when the money laundering occurred, despite the fact that the numbers showed that 99% of the branch profits came from 'non-resident accounts'. This should have raised alarm bells despite the language gap (Bullough, Moneyland—Why thieves & crooks now rule the world & how to take it back, 2019).

Howard Wilkinson served as branch manager between 2007 and 2014 and warned its supervisors in late 2013 about compliance breaches, and he concretely mentioned that some of the clients could likely be linked to the Putin family and the FSB, the Russian Federal Security Service (Milne, 2018). Nevertheless, it took many more years before the scandal reached global attention.

In 2014, Wilkinson resigned because his warnings were ignored. When the news broke in 2018, the bank admitted that the branch moved rubbles out of Russia, converted them into dollars, and moved them to the United States through Bank of America, J.P. Morgan, and Deutsche Bank (Schepis, 2022).

You can imagine how the bank's advisors had to deal with a lot of negative feedback and emotional questions from their clients. It was not only the anti-money laundering that affected the bank's reputation. Additionally, the bank was facing a scandal regarding its Flexinvest Fri investment product,[1] where it was found that the bank had been overcharging for those products compared to the expected return. Evidently, the fraud case was only one example of a bigger problem within the bank.

The fraudulent activities clearly damaged the bank's reputation, leading to the resignation of several executives, including the CEO. The fact that the bank had its nationality in its name made the situation even worse, as the brand was seen as a reflection of the Danish identity.

Chris Vogelzang, then CEO, shared in the Financial Times: "Our customers must be able to trust that we have their best interests in mind and that we provide correct and adequate information and advisory services to them. We have not lived up to this in this case" (Khalaf, 2019). The quote makes his departure 18 months later even more painful after being accused in an anti-money laundering case at ABN AMRO[2] (Skydsgaard & Anthony, 2021). Chairman Gerrit Zalm also resigned because of the ABN AMRO scandal.

5.2.2 A Sustainable Transformation to Rise Like a Phoenix from the Ashes

Carsten Egeriis took over as CEO (Danske Bank, 2021).

It had become evident that something needed to change quickly. The bank's reputation had hit new lows, and people's trust in the organisation was following the same direction. As we all know, trust is crucial for a bank's survival, but rebuilding it is a long and arduous process that requires a lot of effort, time, and a consistent narrative. Danske Bank recognised this and set out to redefine its values.

With a series of changes in the leadership, the bank's culture also started to shift. The bank recognised that financial crime is a key priority and set new measures to avoid any failures in the future (Danske Bank, 2022). I spoke to some of the executives and learned that they changed the entire executive committee, which led to a new vibe in the organisation. They have implemented a purpose-driven governance focusing on sustainable transition, robust and resilient banking, and social impact management. They rewrote their story, reconsidered their core values, and redefined their strategies to transform the organisation in a meaningful way by embracing ethics and values. The bank established a new management, re-evaluated its governance, and refined its purpose: "We release the potential in people and business by using the power of finance to create sustainable progress today and for generations to come" (Danske Bank, 2024).

Their latest Climate Action Plan Progress Report of 2023 (published February 4, 2024) is 85 pages long, around 30 pages longer than the first report a year before. Without going deep into the conclusions of the report, it is impressive how far Danske Bank had gone in creating transparency in its Green Transition early on. With less than 0.01% of its measured emissions coming from its own operations, Danske Bank clearly sees its role in the Green Transition for 99.99% in the lending and investment activities. "The responsibility for the execution of the Climate Action Plan is placed within our business units to ensure that actions on sustainability are embedded within our core business", the report says. The CEO chairs strategic execution oversight (Dankse Bank Group, 2024).

From what I was told, sustainability is not just a box-ticking exercise. They are serious about this and respect the principle of full transparency. The Climate Action Plan Progress Report is a good example. They are not interested in adding a flashy carbon tracker to their mobile app because they know that the biggest impact a bank can have, is by guiding clients to more renovation projects and a transition path to net zero.

In that respect, it is worth noting that the bank also decided to end the customer relationship with ExxonMobil. The reason: "Lack of progress in the company's contribution to the sustainability area, specifically including questionable lobbying methods that, according to our analysis, a) lacked substantially in terms of transparency while also b) working against the green transformation of society". That does not mean that they exclude all oil and gas companies, but there must be confidence that the supplier will commit to a transition that is measurable and concrete (Dankse Bank Group, 2024).

The bank is taking significant steps to steer its clients towards a more sustainable future. While it continues to invest in some oil companies, it has also stopped investing in others that failed to deliver on their next steps to achieving net-zero emissions in the foreseeable future. Additionally, Danske Bank is actively committed to promoting biodiversity and human rights, ensuring it covers the full ESG spectrum.

5.2.3 Learning About Ethos the Hard Way

I spoke to one of their executives, who shared that the bank isn't investing massive amounts of money in a marketing campaign to polish its reputation. Instead, they use the money for more tangible investments, like the green transition and digital transformation. They believe that a great digital experience, in combination with good advice from the bank's advisors, is critical to their high NPS score. Yet, the bank's reputation lags behind that NPS score.

How can a bank rewrite its narrative, redefine its ethos, or, in Danske Bank's case, shift the perception of "banking at a bad bank with a good advisor" into "a great bank in the digital world"?

Dankse Bank may have drawn a line under the recent dark chapters of its +150-year history, but it still needs to prove that it is living the narrative it wrote. The scandals are still challenging them today to effectively rewrite their ethos. As explained in the previous chapter, ethos belongs to the audience; in that respect, a change in the bank, as big as that change may be, is not enough. The change must be consistent and sustained over a long period to enable people to gradually understand that the bank is committed to creating a better planet and society. This is a multi-year project that requires a long-term vision.

Dankse Bank is a great example of an institution that understood that things needed to change. As a national monument, it is more than a bank. It cannot ignore its social contract, which is what happened in the past. The bank ignored the basic principle of good banking in a seemingly small and unimportant but very lucrative branch. It decided to accept the profits without asking many questions. Standard compliance rules were ignored (they didn't apply the principles of logos), leading to (one of) the biggest money laundering cases in history. That affected trust and reputation: people lost their confidence (they lost their pathos), and consequently, they couldn't convince the audience that all problems were a thing of the past (lack of ethos).

The bank is now in the healing process of redefining its ethos, working step by step until society and its people embrace it again as the national symbol it used to be: the engine for globalisation and industrialisation in the past and the engine for a Green Transition today. They understand that talking about the right things isn't enough, which is why they put all their energy into changing the direction of the big tanker that Danske Bank is to make their stakeholder feel that they are serious about their intentions.

5.3 The Greek Drama

Let's go back to the roots of Ethos: Greece, and use the Greek debt crisis to explain the interplay between ethos—pathos and logos. One of my friends reached out to me because he had a story to share. It was Alexandros

Efstratoglou, who was the business development director at Relational FS back then. He asked me: "Why are there 3 big international Greek companies in the software business of non-performing loans (NPL)?" He had my attention.

5.3.1 How the Greek Banking Sector Lost Its Ethos

Many of us still remember how much Greece suffered in the years following the Global Financial Crisis due to various reasons. One of the reasons was that, despite the European Union's target to reach government debt levels of 60% by 2000, Greece's debt-to-GDP ratio remained close to 100% of GDP. Greece joined the European Monetary Union in 2001, which brought a period of high growth and low-interest rates. However, despite these favourable conditions, the Greek government saw its revenues fall and expenditures rise, causing the debt-to-GDP ratio to remain above 100% of GDP.

The Greek government increasingly relied on government bonds and additional debt to cover expenses. This weakened Greece's fiscal position to such an extent that the country was not sufficiently resilient to absorb a higher deficit to support a contracting economy. It led to dramatic declines in GDP. As a result, the Greek government bonds were downgraded to junk status in 2010 (Sarafidis et al., 2017).

The Greek crisis, as it was called, led to impoverishment and loss of income and property. Well-educated Greeks fled the country, and social exclusion increased, all because of the sudden reforms that came with bailouts from the European Commission, European Central Bank, and the IMF.

Returning to Alexandros' question of where there are three big international Greek companies involved in building NPL software, the Greek crisis also led to 45%–50% of non-performing loans in 2010 in his country, he explained: 150 billion euros. 1 out of 2 loans defaulted in the country in 2010. Next to the obvious reasons for the loss of jobs and reduction of salaries because of the economic meltdown, Alexandros also referred to the lack of regulation, people's mentality, and the bank's business model. He explained that until then, banks stimulated consumption by actively promoting credit cards and consumer loans to everyone without strict credit policies (Coeckelbergs, The Greek credit tragedy and the new ethos in banking, 2023).

5.3.2 Credit Management as an Export Product

The issue with Greek banks was not solely related to consumer lending. These banks were significant holders of Greek debt. As a result of the restructuring process, they were required to write off substantial amounts of debt, necessitating government recapitalisation to avoid bank failures. Their financial situation made it impossible to raise capital through private markets.

So now banks were owned by the government, which was under supervision itself. Greece had to regain control over its finances, with a focus on cost rationalisation. Austerity made banks very cautious about taking additional risks and thus reluctant to provide more loans to the Greek economy, whether that was the government, companies, or consumers. The focus was/is on "evergreening" existing non-performing loans, hoping for a recovery, with little opportunity to invest in new economic initiatives (Karamichaildou et al., 2017).

Alexandros explained that after the start of the crisis, banks' first NPL auction issuance required 7 years of negotiations and regulatory reforms. Still today, the process of issuing NPLs by banks on the primary market is so cumbersome that it takes 2–3 years after the first unpaid instalment to sell the NPL. Every extra day further reduces the quality of the loan.

Over the past years, players have entered the Greek NPL market. Companies like Credit Management Services Companies, so-called Debt Servicers like Intrum, DoValue, and Veraltis Asset Management, and investors such as Bain Capital and Apollo entered the market to restructure entire NPL portfolios after buying them and selling them back to the banks. Big corporate loans were the preferred portfolios.

These institutions did much more than weigh their clients' financial risks. They investigated refinancing opportunities and bridge loans and evaluated the company's business model and cost structures. Turning around these businesses takes skill and expertise, skills that banks often lack. Alexandros explained that because of these skills, they call these restructurers "corporate artists".

It is a win-win-win scenario: the corporation would improve its financial situation, the Debt Servicers and Investors would make a good profit by restructuring the loans they acquired at a serious discount, and banks would release capital and improve their balance sheets.

To ensure sufficient financing opportunities in a frozen bank market, a new Greek regulation is coming to allow Debt Servicers to become lenders themselves. The future will tell whether that is the right answer to tightening liquidity in the market. These organisations and their corporate artists often target companies that cannot present sufficient proof of creditworthiness at a regular bank (Coeckelbergs, The Greek credit tragedy and the new ethos in banking, 2023).

It creates a shadow banking circuit, which we don't know where it ends. The story made me question the role of banks in Greek society. Because of the economic situation and the involvement of the Greek government, banks seem to be frozen in adding oxygen to the economy as they look for safe assets to feed their deposit business.

15 years after the start of the crisis, Greek banks are finally starting to heal from the trauma of the government recapitalisation and the austerity linked to it. Of 40 banks that existed prior to the crisis, only 4 remained, explained Alexandros.

One reason was the expansion of credit to the private sector in the years before the crisis, but an even bigger cause was their support in facilitating government debt, which was believed to be a safe investment after Greece joined the European Monetary Union.

Banks lost their ethos because the world rightfully lost confidence in the Greek financial industry. The industry was contaminated because many of its assets, which were assumed to be low risk, were suddenly downgraded to junk.

5.3.3 When Banking Stops Being About Realising Dreams

I asked Alexandros how all this relates to ethos and he shared: "I really liked what I heard from the Chairwoman of vdk bank at your conference, that banking is about realising dreams. Whether that is through a mortgage loan, a corporate loan, or something else, they want to drive things forward. That is what ethos is about for me. Banks have, especially in Greece, lost that momentum because they didn't have the respective capital buffers".

Banks overplayed their hand, and suddenly, the common belief in government bonds and the Greek economy at large evaporated. Banks lost their logos. From then on, their only focus was to find a way to reduce their dependence on non-performing loans instead of supporting their clients in realising their dreams or managing their financial well-being.

Stress took over, making banks lose connection with society and leading to an absence of empathy with clients. The focus was on healing from the crisis, on dealing with their own troubles, and less on supporting their clients' healing from the crisis. Greece was in trouble, and banks were in the middle of it, unable to provide support.

Because of that, banks lost their ethos. That trauma is still affecting the banks' business models today to the extent that the Greek government is now adopting regulations to give non-bank institutions a chance to help people realise their dreams by issuing loans themselves through a less regulated circuit. Will this shadow circuit of loan provision be the right strategy going forward?

Banks exist because they transform short-term deposits into long-term loans. If one of those elements falls out of the equation, what is a bank's role?

Alexandros: "I believe the main topic for banks regarding ethos is to make dreams come true. The mission is to continue supporting the real economy with a healthy risk appetite and as a service to the customer. Banks need to go back to their roots. In Greece, the non-performing loans of most big banks have been carved out, and entire departments, including both personnel and assets, have been transferred to the respective Debt Servicers. Banks got rid of most of their NPL. Now it is time for them to play their societal role again, respecting the limits set by the supervisor and realise the dreams of their clients".

After the interview, a few press releases showed that Greek banks are slowly recovering, with the state-controlled Hellenic Financial Stability Fund (HFSF) deciding to start selling some of their stake in National Bank, Eurobank, and Alpha Bank (Reuters, 2023).

The Greek banking sector is an excellent example of an industry that became paralysed in the face of survival challenges. This was an industry that had lost the trust of its customers and lacked the energy, the funds, and the trust of shareholders to reinvent itself to regain that trust. The focus was solely on survival, and it became an individual process with little room for empathy or concern for others.

It made banks very cautious of their responsibility as lenders for growth and prosperity. That opened doors for new, less regulated players to fill that role. Banking stopped being about realising dreams for people and society and was left available for only the low-risk happy few.

Before redefining their ethos, these banks will need to crawl out of the valley and recover. Only then can they start focusing again on Pathos, understanding what challenges their customers and prospects face and how to help them. Only then can they slowly redefine their ethos and transform themselves meaningfully with a clear purpose and corporate values.

It is a work in progress today. After my conversation with Alexandros, I read some articles from the Greek banking sector with good news. The Hellenic Financial Stability Fund is accelerating its efforts to sell its shares in banks as the market is becoming more attractive again. At the time of writing, Greek banks were expected to pay dividends again in 2024, for the first time since 2010 (Reuters, 2023). In July 2024, Greece even welcomed another neobank, Snappi that was granted a licence from the European Central Bank: "By strategically positioning ourselves outside the Greek capital, we aim to support the local community, nurture a thriving regional tech ecosystem, and spark innovation from wherever we are" (Snappi Bank, 2024).

Redefining that ethos is a long process. Despite being risk managers and attempting to maintain control over their business and the market banks are active in, things can quickly shift in another direction. Our next example will further explain this.

5.4 The Polish Catastrophe

Since 2008, a big scandal has made headlines in Poland, causing many individuals to lose a significant amount of money. The same scandal also occurred in Hungary, Austria, Croatia, and even in France, although with less media attention. I had never heard of it. However, the case of Swiss-franc-denominated mortgages continues to be brought up every once in a while, even today. In January 2024, for example, BNP Paribas announced that they had settled for 600 million euros (Rosemain, 2024).

5.4.1 The Polish Stairway to Heaven

For a client, I was asked to speak about the ethos in banking in Poland in October 2023. After my talk, a few people asked for my opinion on an exotic lending product. I understood the basics, but I became curious and scheduled a call with a banker from Poland who was closely involved in this product (and preferred to remain anonymous). He has been involved in the issue for many years and explained it in a nuanced way.

In Poland, owning a house was considered a status symbol, he explained, indicating success in the new capitalist economy after the fall of the Wall in 1989. Renting a house, on the other hand, was considered a temporary situation for students or at the start of a professional career. Banks were eager to help people fulfil their dream of buying a house and offered two options: a regular mortgage or an FX mortgage.

The FX mortgage served as a valuable financial product that assisted numerous Polish consumers in purchasing their homes. The reason is simple: the interest rate in other countries was lower than the local Polish rates, allowing for more attractive mortgage conditions to realise people's dreams. This, in turn, made this mortgage more affordable for a wider range of people due to its favourable conditions. As an example, in 2008, the annual interest rate in Zlotys was around 8.7%, compared to 4.4% in Swiss francs (Harper, 2023).

Various currencies, such as the Deutsche Mark, euro, US dollars, and Japanese Yen, were used in the past, as they were considered stable and secure economies. However, the Swiss franc was the most commonly used currency, which is why the incident is often referred to as the "FX Mortgage case" or, more specifically, "Poland's Swiss franc case".

The FX mortgage became a big business in 2004 when Poland joined the European Union. Hundreds of thousands of Poles started taking mortgages in foreign exchange, mainly Swiss Franc, anticipating a bright future and steep economic growth (Wlodarczak-semcwuk, 2023).

The country was in full development, and in 2004, the future looked bright. The general consensus was that the Polish economy was on an upward trajectory and would surpass the growth of more established economies like Switzerland. Although banks warned consumers about the risks involved, the perceived risk was low, and people believed that the chances of making a profit were higher than incurring a loss.

The government encouraged the development of foreign exchange (FX) mortgages, as Poland had recently joined the European Union and was one of the fastest-growing economies in Europe. The government saw it as a way to improve their economy through real estate growth. On top of that, they made themselves popular with their voters, allowing more people to own a house thanks to new stimuli.

5.4.2 And Its Highway to Hell

Enter 2008: the fall of Lehman Brothers made people slide nervously on their chairs. The financial world and the world at large were shaking, and the source of all evil was real estate. Although the causes of the financial crisis were American, the parallel with the Polish financial services market was quickly made. Banks were sitting on an enormous pile of FX positions and were trapped in equity gaps, with those mortgages being much more expensive than the underlying claim.

Even after selling the house, people couldn't repay the mortgage in this new economic reality. Close to 700,000 FX mortgage loans were in circulation at the time, worth \$31 billion, 30% of all Polish mortgages (Goclowski, 2019).

The Polish Financial Supervision Authority (known locally as KNF), the Polish banking sector regulator, stepped in and published recommendations to limit the origination of FX loans to Polish customers to those who earn their salary in the respective currency or those who had enough funds to pay off the loan. That limited the foreign exchange risk for consumers who signed up for new mortgage loans. Existing mortgage contracts remained at risk.

The crisis remained controlled as long as people continued paying their mortgages. Unfortunately, the impact of the financial crisis was only a teaser for what was to come.

On January 15, 2015, Switzerland made an announcement that they were going to abandon their currency peg with euro of 1.20 to the euro. This caused a 20% jump in the Swiss franc. As a result, overnight, the Polish zloty to Swiss franc exchange rate jumped from 3.5 to 5 Zloty. The argument of holding a mortgage in a stable and safe environment was no longer applicable because only FX mortgages in Swiss franc were problematic. The other currencies did not generate such a significant loss over such a short period (Goclowski, 2019).

Around 20% of the victims revolted against their banks due to societal drama, leading them to file claims against their banks. However, Polish courts decided that banks had warned their customers about the potential FX risk sufficiently. The main reason behind the customers' revolt was the lack of transparency in how the FX rates were calculated. But the court concluded that banks used a market standard by calculating a daily FX rate, which should have been sufficient. Despite this, around 150,000 customers filed complaints. The high cost of hiring a lawyer, which could easily cost 4–5,000€, was the reason why not all Polish customers filed a court case.

This group of bank clients in Poland took their case to the European Court of Justice, claiming that customer protection was inadequate. The court agreed with them, and since then, there have been more than 130,000 legal cases in Polish courts, with customers winning 99% of them. The verdict was "contract invalidation", as if the mortgage transaction had never happened. As a result, borrowers were to be refunded all the payments they had already made and the bank was expected to receive the initial value of the loan principal.

It left scars on the Polish banking sector. Bear in mind that, for example, MBank had 51.6% of its mortgages in Swiss-denominated loans, Bank Millennium 41.3%, and BNP Paribas Polska 36.9%. Banks like Getin Bank were being restructured in 2022 by a joint venture of a fund set up by the eight biggest banks and the Bank Guarantee Fund (the institution responsible for protecting deposits of private individuals). Capital ratios were below statutory and regulatory thresholds for too long, and they carried a huge portfolio of FX mortgages, which played against them. The bank was being split into a good and a bad bank, where the bad bank purely existed to handle the FX files.

Other banks, like Raiffeisen Poland, Deutsche Bank Poland, and BPH Bank, were put up for sale, but no one risked taking them over as whole because of their FX mortgage portfolio. They were also being split up into a good and a bad bank, with the healthy part being sold to other institutions.

The bad banks required a capital increase to survive and had one purpose: service the portfolio of FX mortgage loans, handle court cases, and try to settle with customers as much as possible before they take the bank to court. In the best case, that leads to a remaining capital that can be paid back to the shareholders. That mission should take another ten or more years, until the termination date of the longest remaining mortgage deals.

5.4.3 Ethos Belongs to the Audience

On June 15, 2023, The European Union's top court ruled that "EU law does not preclude, in the event of the annulment of a mortgage loan agreement vitiated by unfair terms, the consumers from seeking compensation from the bank going beyond reimbursement of the monthly instalments paid", the court said in a statement. "By contrast, it precludes the bank from relying on similar claims against consumers".

The ruling means that Polish banks are facing a $25 billion hit on the ongoing Swiss franc mortgage case. Despite the ruling, the financial regulator stressed that Polish lenders were safe. Even if the ruling would not end up being a systemic risk on the Polish banking market, it does put additional pressure on people's trust in their financial institutions.

Poland was the only country where the product was so widespread and at such a scale.

This case is so interesting because these Polish banks didn't act in bad faith when signing the mortgage contract. Like any other economic actor, they were convinced that the Polish Zloty would grow stronger. They saw that the Swiss franc was pegged to the euro, and all that happened in stable financial times.

They arrived in a perfect storm. But OK, they should have been more cautious. Management boards back then were blinded by the short-term gains of FX mortgages without much consideration of bad experiences in the past, like in Italy. Overall, in 2004, FX mortgages made banks realise more people's dreams. It was with the best intentions but with the wrong risk calculations.

They mismanaged their risks, and as everything collapsed, they pointed to the signed contracts to keep consumers liable for the losses instead of helping people in their financially stressful situations. They sold it as a safe and lucrative product and came back as a boomerang.

They lost the trust of people, politics, and society, the same people who accepted the risks in the first place, wrongfully assuming that the repayment conditions of the mortgage could only improve in the future.

Logos, pathos, ethos...

5.5 How the Fall of Silicon Valley Bank Sent a Shockwave Through the Entire Industry

There are different types of banks; some are universal banks, some are pure retail banks, others are in the private and wealth space or corporate banking, and some are niche players. However, there are also banks that fall in between these two categories, too large to be called a niche but too specific to be called a universal bank. Silicon Valley Bank (SVB), established in 1983, was such a bank. It targeted the tech industry and provided services to nearly half of all US venture-backed technology and life science companies. Additionally, it served many more in countries like Canada, China, Denmark, Germany, the United Kingdom, and many others.

The low borrowing costs and the rise of new companies serving the digital economy after the pandemic had made technology companies very attractive to investors. The focus on growth over profit had led to a huge influx of capital, which needed to be managed by a financial institution.

Over the past 40 years, the bank has established its business in the hub of technological innovation, Silicon Valley. As a result, SVB had immensely benefited from the trend of investing in technology companies. Its assets had more than tripled from $71 billion at the end of 2019 to a staggering $220 billion by the end of March 2022. Additionally, deposits had grown from $62 billion to $198 billion within the same period (Ziady, 2023). The bank relied significantly on the network of start-up companies to assess creditworthiness and risk. The bank favoured both venture capitalists who invested in start-ups and the start-ups themselves as their clients. Even if a start-up founder did not have a salary, they could still obtain a mortgage as long as a trusted venture capitalist backed the start-up. Also, the other way around, most start-up companies blindly trusted SVB to manage all their money instead of working with multiple banks.

SVB had a clear ethos: it was trusted by an important community with lots of money. For 40 years, it built an impressive business around the very specific needs of its customers, based on a solid understanding of those needs (pathos). That understanding was translated into products and services tailored to those customers, making them very loyal to the SVB brand. That led to a pull effect of customers bringing in new customers and deposits. It was a great narrative for as long as it worked. These assets were invested in low-interest government bonds.

Unfortunately, the bank made one mistake: they misunderstood the importance of logos and the rational reasoning behind numbers and risks.

After the pandemic, the world normalised its digital footprint, and because of rising inflation, interest rates went up. That heavily affected the tech industry, where suddenly, pure growth was no longer enough to convince investors. Tens of thousands of workers in the industry were fired, and private equity players pulled back from investments. Start-ups pulled deposits from SVB as they were burning cash to stay alive at a time when SVB's investments were suffering from the higher interest rates, raising rumours about financial troubles. Low-interest government bonds needed to be sold at big discounts (Vynck, 2024).

On March 8, 2023, the bank announced a capital increase at a serious discount to ensure its obligations to its clients, who pulled more deposits. That triggered a domino effect that led to the bank's collapse two days later.

On March 9, a total of $42 billion was withdrawn from the bank. A few influential investors expressed their concerns on Twitter, which quickly spread to other social media platforms and group chats. This led to a bank run unlike any other seen before. It is the first time people pointed at social media to explain the speed at which a bank run happened, and how digital banking allowed for a much faster response to bad news.

March 10, 2023, was a critical date for the entire banking industry. It was the day that Silicon Valley Bank collapsed. Despite strong opposition, the government took over the bank. The rationale of the opposition was that taxpayers should not be burdened with saving wealthy and often arrogant tech leaders. However, the fear of other banks failing and causing further economic damage ultimately weighed heavier in the decision-making process (Vynck, 2024).

Two other banks quickly followed: Signature Bank and First Republic. Signature Bank was known for supporting the cryptocurrency sector, and First Republic had many clients in the start-up scene. The American activities of Silicon Valley Bank were taken over by First Citizens Bank. HSBC took over the UK arm for £1 to avoid leaving tech companies with a minimum deposit guarantee. Signature Bank, or most parts of it, were taken over by New York Community Bank, and First Republic is now part of J.P. Morgan (Copeland, 2024).

These failures led to a much bigger fear of a new global financial crisis. Luckily, the regulators and the industry responded quickly, preventing a domino effect. However, it did stir nervousness against financial services again. Start-ups couldn't access their money, and the advice for the future was (is) to spread the risks and not trust any organisation with all your money. Another lesson was to make sure not to rely only on the risk management of your financial institutions, but as a start-up, build your own risk management and ensure a level of resiliency that allows you to continue in the worst case that your main banks go down (Alikhan, 2024).

The story is a good example of how a bank that does not have its risks under control, one that lacks the rational reasoning and calculation to prove it is healthy, goes down faster than it can make a recalculation. Through social media and 24/7 instant payments, people and organisations will drop you faster than ever, which could lead to a bank run in a fraction of the time.

Losing logos in banking will lead to the immediate erosion of pathos, breaking the ethos. The CEO of SVB could no longer convince the world that, thanks to a capital increase, they had their risks back in order, and two days later, the bank was finished.

5.6 Conclusion

This chapter's goal was to show how complex a good understanding of the rhetoric triangle can be for banks and how intertwined these elements are. Because of its complexity and multitude of stakeholders, banks have a limited toolbox at their disposal, and the tools they have require a high skillset to write a new story and redefine their ethos.

Banks are the veins that let money flow across the globe, and they affect everything as everything can affect them. That requires conscious management of uncertainties because many crises start with the best intentions but ultimately have a devastating impact on people's and organisations' trust in their banks.

The emergence of big universal banks meant that the reputation of a few can have a devastating impact on the trust of an entire industry. It can take many years for the industry to recover from such a downfall. In that respect, Danske Bank was fortunate in a way that it lost the faith of its customers due to its own actions, and thus, the responsibility to earn that faith again was solely theirs. They rewrote their story, reconsidered their core values, and redefined their strategies to transform the organisation in a meaningful way by embracing ethics and values. The only thing left now is the time to prove that they are serious about this transformation. Actions speak more than words, and the future will tell whether the actions suffice to make Danske Bank the Danish symbol again that it used to be.

Greek banks faced a completely different challenge. Historically, they had invested heavily in Greek government bonds, which seemed like a good decision when Greece joined the European Union. However, the government could not manage its finances effectively, and the impact of the Global Financial Crisis on the Greek economy made matters even worse. Although the rate of non-performing loans on consumer loans was extremely high (45%–50%), it wasn't the banking industry's biggest concern. The real problem was the significant deterioration of government bonds. As a result, the Greek government had to recapitalise the banks that held these bonds, making it a shareholder of their financier. This paralysed the banks, as they didn't have the capabilities to provide much more loans to the economy. They were unable to fulfil their societal promise of helping people achieve their dreams. The miscalculations (loss of logos) put banks' survival modus (loss of pathos), that way completely eroding the trust their audience once used to back (loss of ethos) at a time when their clients needed them the most.

Luckily, the industry is slowly recovering, and hopefully, soon, they can review their narrative in a more open and constructive internal and external dialogue. This will allow them to claim their position again as a financial intermediary, a role that is now increasingly taken over by these so-called Debt Servicers.

We explained a somehow similar story in Poland, where the early days in the European Union saw strong economic growth and a government that favoured house ownership. With the best intentions and some self-interest, Polish banks creatively created new financial products to grow the pool of prospects for a mortgage loan. The FX mortgages made mortgages better priced, allowing more people to sign for a mortgage. Unfortunately, the dreams that banks realised turned into a nightmare for many. The FX mortgages were a societal drama with many actors to blame, but in Poland, society blamed banks for all problems. None of their calculations and contracts were convincing to prove the opposite. They lost their logos.

After many lawsuits (banks losing connection with their customers, loss of pathos), the banks are now slowly recovering from the damage done to leave the dark page in Polish bank history behind them to write a new chapter in redefining their ethos.

The case of SVB is interesting because of the speed at which a bank can lose its ethos. Banks must always have their risks under control, and in the age of social media and real-time interconnectivity, the rhetoric triangle can break faster than ever. The fear of a bigger financial crisis after the collapse of SVB also shows the fragility of ethos for every single bank because banks are so interlinked.

Ethos belongs to the audience. It is not something you control, and it is very fragile. Ethos is about how the audience looks at you, and in the case of a bank, that audience can be customers, governments, supervisors, regulators, other banks, and so on.

If a bank wants to build a new ethos, all these stakeholders need to be taken into account, or the exercise will not work. Ethos is much more than complying with a set of rules and regulations, as it is much more than preaching the right thing.

Martin Rohner, Executive Director at the Global Alliance for Banking on Values (GABV), once told me: "Every bank has values that define its decisions and actions. If these values have not been clearly defined and reflected, then organisations resort to profit or profit maximisation as the default value. That works if it is a real economy business, but in banking, it can create different challenges. Banks play an infrastructure type of role in the sense that they are enablers of economic activity. An excessive focus of banks on profit can lead to adverse outcomes that can affect the wider society. Certain businesses or segments of the population may not have access to finance, and therefore, this creates inequalities. They may put the economy in jeopardy because they prompt a financial crisis".

Danske Bank grew and lost control over its size, the Greek banks grew and lost control over the values of their core investments, the Polish banks sold enormous amounts of mortgages that they could not have sold in a local currency, and SVB lost control over its balance sheet because the circumstances of the tech and start-up industry changed.

They all bet high and lost their ethos. Some have some energy left to build a new ethos; for others, it is too late.

Notes

1. Flexinvest Fri was an investment product that Danske bank sold and advised to its customers, despite their knowledge that it would not lead to positive returns in the coming 5 years, at fees that were too high. 87,000 customers were affected (Danske Bank, 2019).

2. Chris Vogelzang held many executive positions at ABN AMRO between 2009 and 2016. The anti-money laundering case against ABN AMRO Bank was settled for 480 million euro.

6

Don't Hide Behind Regulation

6.1 Introduction

This chapter explains why ethos is not about having good rules and laws. Ethos is not about a common set of guidelines; it is not about applying general mandatory principles. Ethos sets your strategy of playing the game in a good way and is aligned with general mandatory principles. These laws and regulations can sometimes even limit organisations from defining their ethos in banks. Then again, these regulations exist for a reason, so I am not against them. We need structure and guidance, especially for an industry that has such an immense impact on every person's and organisation's life.

"Regulation and supervision arrive when trust is lost. It is a reactive societal movement to put things back in control", explained Dinesen (Dinesen, Absent management in banking—How banks fail and cause financial crisis, 2020). Indeed, new rules are often the result of the bad behaviour of one that affects an entire industry or the result of an event that made regulators doubt whether the industry had everything under control. Following a crisis, the industry faces new regulations. However, it is worth considering whether regulations alone can restore trust in the industry. Do new directives and restrictions really make the industry more responsible, and does society believe it makes the industry more responsible? Or is there a need for a greater shift in the ethos of banking as a whole?

This chapter argues that the new ethos should be a proactive evolution, with trust at the centre. As much as regulations are based on ethical foundations and logic set stricter rules, it takes more for banks to rewrite their narrative. Regulations create a one-size-fits-all set of measures for the industry,

R. Coeckelbergs, *A New Ethos in Banking*,
https://doi.org/10.1007/978-3-031-78234-3_6

but we learned from the previous chapter that ethos is highly personal and unique for every organisation. As we learned, ethos is very contextual, but not just that; ethos describes our personal, corporate, and even societal convictions and considerations. It is an interpretation of what is good and wrong, contrary to regulations, that looks for common denominators and rules for organisations and people to play the game.

The Netherlands provides a great example of how the industry acted in good faith, with the bankers' oath: a commitment from banks to do good going forward. Many years later, in 2024, Belgium also picked this up and set up a bankers' oath.

Regulations create a level playing field and a framework to act in good faith, and each actor (or association) must determine its strategy in that game. Regulation sets the bar, but organisations must set their own target and their own strategy to reach the right results.

Regulations are put in place to provide banks with guidelines and ensure they do not misuse their power but rather operate with the intention of doing good. These rules set the framework for a bank's operations. However, as we will discover in this chapter, these regulations can also restrict the bank's operations, not always in favour of the positive role these banks wish to play in society.

In this chapter, we will discover that a bank's responsibility extends beyond simply following rules and regulations. Ethos is an embodiment of ethical behaviour within their own organisation, and following the rules is just the bare minimum. The rules are intended to improve and secure the financial industry. However, as they are universal and do not always account for context, they can restrict a bank from defining its ethos and limit its ability to contribute to a more equal and financially healthy society.

6.2 Regulation Overload: Why More Rules May Not Prevent Banking Crises

At the start of banking, people put their trust in a family, and they trusted them with their money because of their values. Today, the industry is different and much bigger. With its size came much more regulation, and rightfully so. It makes banking more accountable, with improved perceived trust in the industry. Although there is increasing pressure from regulators to reduce risks, no one should ignore the role of a bank in society, and that is risk management.

It is clear that regulations are necessary to ensure businesses operate in the public interest. Regulations bridge the asymmetric information, as we cannot know everything. They establish a framework to trust organisations and people in their good behaviour. In this respect, regulations set the rules of the game to provide consumers with the right level of protection and certain guarantees of financial stability to avoid new financial crises. Regulation is required to urge financial institutions to step up their game to fight financial crime, and it also stimulates competition and innovation.

Overregulation may eventually stop serving the public good, so balance it well. That balance must consider customer protection, a good understanding of the industry, and avoiding excess regulation as it affects both the industry's efficiency and competition, which are key to a healthy market economy. Additionally, regulations alone are insufficient for a bank to define its ethos.

There are many approaches to examining regulation. The public-interest approach would say that we need regulation to ensure that the market economy produces outcomes that are in the interest of the customer, be it a depositor or a borrower. In other words, we cannot assume that someone will always act in the best interests of all parties involved. We need rules to prevent issues such as monopoly rent and asymmetric information and to ensure proper customer protection.

On the other hand, there is the self-interest approach, which claims that regulations are set for the banks' benefit. Often, banking scholars use both the self-interest and the public-interest approach to explain the appearance of certain rules. Rules are set up to serve customer protection, but by setting new rules, new entrants will also face a tougher time entering new markets. As much as these rules burden existing players, they also set fences for new players to enter, thus reducing competition and indirectly protecting the incumbent players.

Another approach is the theory of psychological attraction, which argues that participants in the political process have psychological biases that are exploited by the regulatory process. They tend to respond to personal stories that tap into people's emotions—negative emotions more than positive ones, by the way. This could explain why we often see more regulations put in place after every banking crisis. However, not all regulations will lead to the desired positive effects.

Scapegoating is a common practice in the regulatory process, where someone or something is unfairly blamed for problems or negative outcomes to divert attention from the real causes. Hendrickson refers to another psychological process, the one of overconfidence, which is the belief that one's capacity is higher than it actually is, often referring to politicians who vote for new rules and regulations after financial crises, not truly understanding the root cause, nor the impact of new regulations (Hendrickson, 2011, pp. 10–14). This is something I thought about a lot while writing the story about the Swiss mortgage loans in Poland.

Crises are of all times. Sometimes, they happen because of changed circumstances or loopholes in risk management or regulatory framework; sometimes, they happen because of bad intentions and sometimes because of stupidity. Most of the time, though, banks are not the only ones to blame; a crisis occurs because of a combination of factors. Hendrickson (Hendrickson, 2011, pp. 5–7) found four elements that can be found across all bank panics or crises:

- An external event has caused a surge of positivity in both the financial and real sectors of the economy, leading to high expectations for future profits.
- This leads to an increase in credit issuance to those who want to capitalise on the wave of positivity. Banks see loan extensions as less risky and are happy to provide them.
- Suddenly, optimism gave way to caution and a loss of confidence.
- That change in expectations leads to banks failing to fulfil their commitments: "Borrowers realise that their indebtedness is too large, and lenders recognise that their loans carry too much risk".

One could point to a bank's risk management to ensure banks never reach the third element, but reality is often much more complicated. Even in the case of the Global Financial Crisis, there are reasons to believe that banks were not the only ones to blame. Who would have thought at the beginning of the 1990s that the new laws and regulations to stimulate home ownership would lead to a global economic meltdown? Fanny Mae and Freddy Mac were two Government-Sponsored Enterprises (GSE) that purchased mortgages from banks to allow these banks to issue more mortgages. In line with the government's mission to stimulate homeownership, many mortgages were subprime, issued to people with a poor credit history, for example. More people got approved for a mortgage, and the Federal Reserve's low-interest policy reduced the cost of borrowing. This combination of factors led to a steep rise in mortgage demand.

These GSEs could hold the mortgages and earn interest on them, or they would repackage them in mortgage-backed securities, promising investors to pay the interest and principal even if they made late payments. Mortgages were transformed into mortgage-based securities and collateralised debt obligations, leading to a growing shadow banking market that poisoned the entire industry.

The creation of increasingly more complex and lucrative products, triggered by public policies and laws to facilitate home ownership, led to the Global Financial Crisis, which began when the bubble burst in 2007 (Hendrickson, 2011, pp. 205–229).

In Europe alone, 6,7 million jobs were gone between 2008 and 2013. Investments went down by 20% overall (European Parlaiment, 2019).

People were angry and looked for someone to blame. It is an excellent example of scapegoating because, as we explained, the GFC was a disaster with foundations in the early '90s, 3–4 government periods earlier. The Global Financial Crisis was traced back to the untransparent and highly complex financial products created and traded by banks. What was ignored was that the high amount of non-performing loans was the consequence of government measures years before and something supervised by an independent body set up for the stability of the financial industry.

Banks were being saved with government money, which could have been used for more productive objectives, especially in a financial and global economic crisis. As a result, the banks were being criticised for accepting government funds without the consent of public opinion. That public opinion was also struggling from a massive economic crisis and didn't receive the same level of government assistance, which led to envy towards the financial industry, not knowing what the impact on their lives would have been without those bailouts.

In order to prevent similar incidents from happening again, the regulatory body intervened and established new regulations regarding capital requirements, supervision, heightened competition, customer transparency, and ethical conduct. Additionally, higher deposit insurance was made available, although studies have demonstrated that this may equally increase banks' moral hazard. Some argue that deposit insurance assures individuals that their money is secure, even in the event of financial difficulties at a bank, which may make them less critical of whether they choose to deposit their funds at a cautious or risky bank. In other words, clients are not stimulated to be critical of the bank they work with, as there is always a safety net in case a bank goes bankrupt. On top of that, clients will be compensated by an external insurance scheme covered by the entire industry, meaning the prudent banks will be held equally responsible as the less cautious banks (Hendrickson, 2011, p. 66).

The combination of measures prevents banks from falling down again, but it also limits their possibilities of fulfilling their role in society, as we will read later in this chapter. In addition, we learned that additional rules and measures can also reduce competition. Some would even argue that the high number of new rules and regulations is potentially sparking a new crisis in the future because banks are looking for creative ways to find new revenues to compensate for the extra regulatory costs.

In short, regulation is needed, but it needs to be balanced. Regulation should not be purely driven by emotions, because there is too much at stake. Too much regulation will hamper a bank's ability to fulfil its social responsibilities, but the opposite is also true: insufficient regulation risks banks not properly fulfilling their social responsibilities.

"Getting more freedom in the playing field you're in means that you get more responsibility. And having more responsibility means knowing your position in the economy and society and how to deal with it", one interviewee said. That is exactly what ethos is about and why regulation isn't enough. Regulation is needed to ensure the bad apples cannot do much harm, but the ethos is how financial institutions translate their interpretation of responsibility, accountability, and a healthy governance structure, not just in words but even more in their day-to-day operations.

6.3 The Dutch Banker's Oath that Redefined Ethos

I would like to clarify that I did not intend to suggest that banks were not responsible for any part of the financial crisis. In fact, as we read in Chapter 1, the blaming and shaming after the GFC may have been the catalyst for change for a multitude of reasons. Yes, there are valid concerns that banks are too big to fail; yes, banks have become greedy, prioritising shareholder value instead of customer well-being; and yes, there clearly was an issue in fighting financial crime.

The banks were an easy target to blame for everything that went wrong due to the erosion of trust caused by a combination of factors. In that respect, I like what Eelco Dubbeling, Managing Director at the Dutch Banking Association, shared with me: "I recall starting my career in a bank where the focus was on globalisation and consolidation, with a leadership that lived a vision that only a few global banks would remain in the end, with a culture that was primarily centered around career and financial success. The crisis led to a recalibration of culture, products and services. In the Netherlands, we held deep discussions on the role of banks in society" (Coeckelbergs, The Banker's Oath and The New Ethos in Banking (part 1 of 2), 2024).

Many of those discussions were held in the Maas Committee, the Advisory Committee on the Future of Banks in the Netherlands, and it led to the advisory report "Restoring Trust", with new recommendations for the banking industry to do better in the future. The report recommended suggestions for risk management and compliance, executive compensation and corporate governance, transparency and accountability, and regulatory and supervisory framework, and all circled around how to tie the industry back to society and its expectations towards the industry. It also led to the Banker's Oath or Banking Code in the Netherlands.

Interestingly, these commendations came from the industry that acknowledged they were wrong and wanted to do better in the future. They were redefining their ethos in banking, well beyond what was required by regulation. Some of their recommendations even stimulated more or different laws to build a healthier and safer banking industry going forward (Adviescommissie Toekomst Banken, 2009).

The Banker's Oath was an intrinsically motivated initiative by the banking industry to raise awareness of bankers' roles and responsibilities and how to behave within that context. The oath was later transposed into national law.

The Dutch Banking Association has a draft oath on their website which, translated in English looks like this (Nederlandse Vereniging van Banken, 2015):

I swear/promise within the limits of my role that I at any time fulfil in the banking sector:

- *that I will perform my function with integrity and care;*
- *that I will make a careful consideration between the interests of all parties involved with the company, namely those of the customers, the shareholders, the employees, and the society in which the company operates;*
- *that in this consideration, I will prioritise the interest of the customer;*
- *that I will act in accordance with the laws, the regulations, and the codes of conduct that apply to me;*
- *that I will keep confidential what has been entrusted to me;*
- *that I will not misuse my knowledge;*
- *that I will present myself as open and accountable and acknowledge my responsibility for society;*
- *that I will strive to maintain and promote confidence in the financial sector.*

So help me God Almighty! / This I declare and promise!

On [date], was at [place] in the presence of [name of the person in whose presence the oath or promise was made], and in the presence of [name of another representative of the company or sector or professional organization] the oath/promise according to the aforementioned formula was made. Also, [name of the involved person] confirms to adhere to the code of conduct and to recognise the enforcement of the code of conduct and the exercise of the powers under the Disciplinary Regulations of the banking sector;
Involved person
[signature of the involved person]

This oath must be taken a maximum of three months after starting a job in the banking industry, and it must be part of a meaningful ceremony for maximum impact. Banks are free to translate that into their organisation (Merz et al., 2021). In that respect, it is a great example of defining ethos.

Eelco Dubbeling shared a few valuable practical examples of that translation. One example was banks that set up workshops to discuss ethical dilemmas with groups of people from every level in the hierarchy. Another example was a bank that asked its employees to publicly demonstrate ethical behaviour in front of clients to make them think more deeply about the values they preach. Lastly, there was a bank that asked its employees to take an oath in front of a mirror in the hallway and write their name on it so that everyone would be reminded of the oath they took and acknowledge that everyone else in the organisation did it too.

Mr. Dubbeling: "We have learned from this experience and are committed to rebuilding trust with our stakeholders. All this led to a recalibration of what ethics truly means. Nowadays, there is a whole new generation of bankers and leaders in the sector who are much more transparent and have a better balance between their emotions and rational thinking".

Today, the responsibility for monitoring compliance with the banker's oath has been transferred to the Stichting Tuchtrecht Banken, a disciplinary court that enforces ethical standards in the banking industry. However, not everyone is pleased with this change. One individual expressed their opinion that the original setup was excellent because it was created by the sector, for the sector, and focused on dialogue and positive reinforcement. This may no longer be the case today since the responsibility for monitoring has been shifted to a "disciplinary court for minor offences that bankers may have already been found guilty of by a labour court".

I understand that point of view, although I also understand that having an independent body overseeing the ethical standards makes the supervision more independent. The beauty of this banker's oath remains the industry's intrinsic motivation to set it up as a sign to society that they act with the best intentions.

It builds on pathos, connecting with society again, not just explaining how they do it with the best intentions, but also listening to what society expects, like doing the oath in front of customers, for example. To ensure continuity, the industry has defined clear rules and guidelines for its stakeholders on what is good and wrong (logos). That way, they redefined their ethos over the years. The Dutch Banking Association saw this translated in a consistent improvement of their Reptrak (Reputation Institute) rating for the banking sector, explained Eelco: "People highly value and appreciate the quality of products and services offered by banks. For banks, like many other companies, the products and services you offer to clients are of utmost importance. If your products are good and clients are happy, you'll have a good reputation" (Coeckelbergs, The Banker's Oath and the new ethos in banking (part 2 of 2), 2024).

On January 15, 2024, the Banker's Oath was also introduced in Belgium for directors, certain credit institution staff members, and agents in banking and investment services. Karel Baert, CEO of Febelfin, said about the banker's oath: "Febelfin is convinced that this will further increase pride in the banking profession and highlight the important societal role it plays. After all, bankers have important economic, social and ethical responsibilities. It can also further boost confidence in the banking sector" (Febelfin, 2024).

The banker's oath is now part of the banking industry's regulatory framework in the Netherlands and Belgium.

6.4 Regulation as a Catalyst for Stronger Banks

The regulatory framework reduced the options for the banking industry, making it a more healthy industry. The implementation of higher capital requirements has made banks more resilient in the event of a crisis. This was evident after the fall of Silicon Valley Bank, which did not lead to an industry-wide crisis but only a few exceptional bankruptcies and required mergers, such as UBS taking over Credit Suisse. It is clear that banks are now better protected against such events.

The new measures over the past 15 years did make the banking industry more resilient overall. Banks devote a lot of time and energy to complying with new regulations on capital requirements, customer protection, improving competitiveness, etc.

Hedwige Nuyens, Managing Director at the International Banking Federation, shared with me: "I'm not going to complain about too much regulation. It is like complaining about too much education, too many medical devices, and too many medical procedures. Regulation is part of our lives and is also the reflection of the complexity and of the world we are dealing with, the growing expectations of people towards society, and the increasing diversity within society. I believe that law and regulation play a crucial role in structuring society".

It's not just financial-services-specific regulations like Basel III, Capital Requirements Directive (CRD), and Regulation (CRR), Single Supervisory Mechanism, MiFID II, PSD2, PSD3, PSR, FIDA, Anti-Money Laundering Directive (AMLD), Bank Recovery and Resolution Directive (BRRD), and more. On top of those, banks also need to comply with regulatory initiatives like the Digital Operational Resilience Act (DORA), the Artificial Intelligence (AI) Act, GDPR, the European Digital Identity Act, and more. Let's all agree that compliance is a serious challenge, but so far, it seems to be under control, and banks are looking for creative ways to make the best of a bad situation.

Open banking may not have led to the expected spark of increased competition by new players, but it drove a wave of innovation in the banking industry. PSD2 was a legal requirement and forced all banks to invest in APIs and share transaction data. It also fuelled the innovative thinking of more possibilities using the same platform. The most innovative banks started thinking about how to make an opportunity from a legal requirement, and that is essentially what I believe open banking is about: it is not a technological evolution but more of a mindset change in the industry.

Open banking and PSD2 triggered banks to find lucrative business models in this new reality. A logical evolution is embedded finance and embedded payments, where banks look for new channels through partners and corporate clients to distribute their products and services and to find new clients in very specific customer journeys.

In Europe, we see the SEPA Payments Account Access Scheme (SPAA), a new scheme based on the same PSD2 platform but with premium APIs that banks offer with a transaction fee, that way looking for a way to pivot an expensive platform into a revenue-generating new service channel.

Similarly, I see and hear more bankers talking about the need to look at regulation as an opportunity rather than a pure cost. Again, you see that banks are put in a corner and have to incur expenses to comply, but they try to find their own way and take the initiative to regain control over the situation.

Another example is the one of Know Your Customer (KYC) and Anti-Money Laundering (AML) practices, where banks are increasingly being given the role of gatekeepers to prevent the flow of illicit funds. They are tasked with knowing their clients to a greater extent than before. The rationale is that combating financial crime requires prevention at the source where illicit funds can be identified and stopped. Money laundering and terrorist financing are not solely banking problems; they are societal challenges involving many stakeholders. Regulatory frameworks are putting banks increasingly in charge of ensuring that no dirty money enters the regular circuit.

On January 30, 2023, The Banking Scene organised a couple of sessions on how regulation is an opportunity to change the rules of engagement, and a common conclusion was that regulation is an opportunity. The KYC screenings in private and wealth management for example, are a required extra administrative burden for a bank advisor and the client. They see it as unproductive time. One of the panellists explained that this sentiment is now changing. On the one hand, technology allows to keep the work to a minimum, and more importantly: "banks have realised that this customer contact is a unique moment to retrieve much more information from the client, which can help to complete the client's profile to improve the financial advice given to that client" (Coeckelbergs, How regulation is affecting the rules of engagement in banking, 2024).

Banks see the same need today in the investment advisory and the required extra questions in respect to the Green Transition. Regulation is changing incredibly fast, and bank advisors can hardly follow. On top of that, customers don't understand why banks ask the extra mandatory questions. In the panel around sustainability, we heard bank executives explain that they understood the importance of the new regulations: "They are more than just compliance checkboxes; they are reshaping the very fabric of retail and cooperative banking, and they make bank's efforts for a greener future more comparable". It is an opportunity to engage with clients to talk about their values and intrinsic personal priorities to help them steer their investments in the right direction.

Today, most banks are still in denial and not always willing to understand the motivations behind the new guidelines and additional complexity, but banks will need to find a new and positive narrative. The engagement that banks have with their clients will drive real positive change in the end. So, banks will need to outgrow the legal jargon and speak in the language of their clients how their money can leave a lasting impact for generations to come.

As one of the panellists shared: “If our own people and, first and foremost, our relationship managers are not convinced by the positive narrative and by the possible positive outcomes of what we’re doing, we have no chance to succeed” (Coeckelbergs, How the sustainable agenda affects the rules of engagement between banks and their clients, 2024).

Again, regulation is pushing banks to go the extra mile. That is the mile required to rewrite their narrative based on the underlying principles of the law while respecting them.

6.5 How Overregulation Can Hinder Redefining the New Ethos in Banking

Regulation translates what a society believes is ethically important within its geographical and societal context. By putting that in a law, organisations and people are forced to adhere. These changes can obviously lead to resistance, but we now know that more proactive organisations try to make the most of it, using the new rules to change their rules of engagement (in the application of the KYC regulation, for example) or to go the extra mile, trying to make a cost into a profit (like with PSD2 and SPAA).

So, regulation is a good thing, despite the limitations for people and organisations. At the other extreme, there is overregulation, where regulations can hinder organisations and people from doing good.

6.5.1 The Opportunity Cost of Compliance

Whether regulations are about higher capital ratios, additional checks and controls, adding new services, or offering basic bank services for clients a bank would rather not want, compliance comes with a cost. In today’s financial landscape, a considerable number of resources are dedicated to compliance rather than being used for other advancements. Consequently, innovation didn’t always go at the pace some people in society expected. Or branch networks were being closed down much faster than some people in society expected. It’s challenging to quantify the opportunity cost of investing in compliance instead of focusing on innovation and improving customer experience. However, it’s evident that these additional investments put pressure on profits and work pressure of employees. Purely on financial crime compliance, the company LexisNexis found that the United States and Canada have an annual cost of $61 bn (PR Newswire, 2024), £34.2 for the UK (Kerkez, 2023).

A study by EY and ABBL showed that in Luxembourg alone, 2019 represented a total cost of 548 million for banks on regulatory matters. More important is the impact of regulation on smaller institutions. The same study showed that the big banks in the country (67% of total assets) represented 52% of regulatory spending, compared to the smaller institutions that covered 2% of assets, has a burden of 6% of regulatory spending (EY & ABBL, 2021). Laws and regulations apply to global universal banks and smaller local banks. Although there are gradations in the function of a bank's size, the pressure from supervisors and additional new rules is immense.

It made banks reconsider their priorities and even figure out how to fund the additional regulatory developments.

Banks look for ways to rationalise the cost of compliance, and partnerships are a valuable option for areas outside the competitive space. Through collaboration with their peers, banks find synergies and spread the costs over multiple organisations, making resources available to shape the bank's mind and heart.

There again, fintech companies come to the rescue or let's call it Regtech, as many of the firms often serve financial and non-financial firms with their solutions. Regtech comprises "a group of companies that employ the use of technology in aiding businesses to comply with regulations more efficiently and in a cost-effective manner" (Vanderhoydonk & van Beek, 2019). These companies assist banks under the hood to keep the cost of compliance to a minimum and are highly successful in it. Not only can they scale their solutions over many banks to reduce the cost, but they also learn from all their implementations and partnerships to build better services than most individual banks could ever do on their own.

Working with these companies helps banks minimise their opportunity costs. Over time, banks have become more willing to outsource work on their core engine to third parties, with compliance being an integral part of this. They recognise that this strategy will reduce the effort required to remain compliant and free up resources for providing value-adding services.

Overall, the conclusion from this part, though, is that implementing new regulations and adhering to them comes at a cost and focus that could be spent on more valuable developments and activities.

6.5.2 Regulatory Limits in Serving People and Society

So yes, regulations are expensive and come at the cost of other investments. Now, let's look at what other consequences regulation can have on redefining a bank's ethos. Redefining your ethos is all about how you play your role as a bank, in the products you offer, the services you provide, and the values you embrace in doing so.

The markets and customer expectations are constantly changing. Therefore, banks need to regularly review their processes to ensure they don't need to redesign their products or services due to these changes. The guidelines on product oversight and governance arrangements for retail banking products, set by the European Banking Authority, have good intentions, but they can impact the speed at which a bank can adapt and affect the resources required to do so. These guidelines establish clear rules for managing the risks of product oversight to ensure that customers are adequately protected in the product development process. It takes away pragmatism and slows product development cycles. As risk management is involved in the entire product, banks will witness reduced flexibility and a higher compliance cost (European Banking Authority, 2015). In developing a new product, banks are much better equipped to face the risks of new products and services. However, based on my personal experience, I know that the additional time-consuming reports that are often required don't add value in most cases. Instead, they seem to be administrative tasks that simply allow product managers to check a box. Bankers who take their job seriously don't need guidelines to understand the potential risks of new products and services and will consider those risks without lengthy procedures as well.

Many new regulations start with the going-in position that banks must be prevented from doing wrong instead of assuming they'll do good. They aim to enhance banks' core resilience and ensure the safety of their customers. These regulations intend to positively affect the rules of engagement, but often the opposite is true because of the top-down approach: "We tell you to do it this way", and change management is less of a priority. While these changes can help prevent banks from engaging in wrongdoing in the long run, their impact on customers' perceptions of their bank is not always easily measurable, but instinctively, it doesn't feel like it is going the right way. In the context of KYC, but also in investment advice for example, regulations will demand from banks to ask their customers the necessary questions, to tick the box, rather than directly affecting their relationship with customers in an empathetic way.

These rules are set with the best intentions to make the industry safer and more resilient. As they target an entire industry, the measures are based on "the average bank" and don't leave much room for organisational interpretations in creating a sound financial environment for clients. Rarely do they distinguish between the bank that treats its client as a number and the one with a strong local anchoring and a good understanding of its client's (financial) situation beyond its assets and transaction history at the bank. With some regulations, this can lead to perverse effects when you think about how banks can and should do good for people and society.

Take the example of IFRS9. That regulation defines how to measure and qualify non-performing loans. It reduced the options for banks to support their clients in financial stress. IFRS9 mandates, for example, to classify a loan with a delay of 90 days in payments into non-performing. That period is not negotiable, irrespective of the context, whereas in the past, banks had more options. They might sit down with the client and evaluate the options. Sometimes, a bit of extra time was enough to solve the problem. No one is interested in non-performing loans: it costs the bank extra money, and the client may lose some of his assets to pay the loans. In that respect, IFRS limits the bank in playing its social role as a guardian of financial health. Moreover, these stricter regulations make banks more reluctant to lend money to people in the higher-risk category, which allows fewer people, for example, to buy a house. "Somehow, the possibility of having more ethos is taken away from the hands of the banks", an executive told me in that context (Coeckelbergs, How regulation is affecting the rules of engagement in banking, 2024).

Another example comes from Beobank, a Belgian bank and part of a bigger cooperative group French Crédit Mutuel, that, when I spoke to them, was investigating how to assist every citizen in the transition towards green living. Their CEO, Guy Schellinck, told me, "I am also convinced that if we keep handling the transition the same way we are doing now, the inequality between the rich and the poor will keep growing. At Beobank, we are striving to find a solution to this challenge, but we acknowledge that it's not an easy task".

At the time of the interview, February 2024, Beobank was investigating with one of the major Belgian energy suppliers how to build a cost-neutral product for the consumer in the short term but leaving a lower footprint in the medium term. As in many other interviews I had, legal restrictions were a constraint in building a product that would let people pay the same amount for money for energy and as that the required investments in a greener future today, as if they did nothing (and kept paying the high energy bill they already have).

In that context, I would also like you to think about the following statement from Roland Van der Vorst, Head of Innovation Wholesale and Rural at Rabobank, who once shared with me: "Traditionally, banks have collateral and look back to historical results to assess their long-term value. In the context of sustainability, however, we see that soil is degrading. If you want to invest in farmland and restore natural capital, you only reap the benefits, probably only after three or four or five years. So, what you need to do there is actually do it the other way around. You need to translate future value into today's benefits. The challenge for these farmers is overcoming this 4–5-year period, the Death Valley".

But what do you do if that is not allowed?

I want to close with one last example of how regulation may have a destabilising effect on redefining the new ethos is banking, more about the general theory of regulation. Hendrickson found evidence that because regulation affects the profit opportunities of a bank by altering cost and revenue structures, it may have a destabilising effect on the industry, making banks more creative in finding new revenue pools that are not affected by new regulations. The off-balance credits in the past are a good example of that (Hendrickson, 2011, pp. 16; 169–170).

6.6 Trust Leaves on Horseback and Comes on Foot

The digital world has redefined the way we think and how we expect things to work. Everything needs to work, there are close to zero emotions involved in the process as long as things go well. "Convenience is a dissatisfier" said a bank executive once, meaning that convenience will rarely make people happy, but it sure will make people unhappy when things go wrong.

Equally, regulation rarely triggers happy emotions that contribute to a deeper connection between the industry and its customers, but when things go wrong, people can point to regulation to explain how bad things were managed, for example. New rules were voted to avoid unexpected negative effects or harm from banks, but they mostly don't create delight from a customer or societal point of view. To an extent, we could even say that regulation may have had an adverse effect, as extra rules and procedures can also be interpreted that banks need more rules to do good instead of leaving room to redefine their own fate.

After a crisis, trust is lost. The world expects action to prevent future crises, so regulators set new rules. It is a reactive move after trust has fallen. Trust can go unexpectedly fast, and Silicon Valley Bank was an excellent example. Regulation will not bring back trust. Even if you think it does, it won't solve the trust issue immediately. Just look at how long it can take to set new rules and procedures. Trust is external, like ethos. Others define how trustworthy you are. After a crisis, it can take a long period of grief before trust is restored.

Regulations may make it easier to restore it, but they will never solve the problem.

Regulation provides stability based on ethical considerations, but that is not what ethos is. Ethos is much more. Ethos defines how a bank interprets the rules. That is where ethos, the organisational interpretation of what is good and wrong, is key. Banks must proactively align their values with their actions and ensure they match societal expectations. What do different generations expect from life, and how do they expect their bank to help with that? Banks must translate these insights into defining their ethos because that is what they will be accountable for in the long run by their clients. How will each bank make its own footprint in bringing delight to customers in a responsible way?

A good regulatory framework, combined with room for banks to play their role, governance, responsibility, and accountability, is essential for rebuilding trust in the long run. Banks need to make sure that they avoid the penalties from supervisors (regulatory framework) and comply with that (unwritten) contract with their clients, and that is essentially why regulation isn't enough for the industry to redefine its ethos. As a customer, you need to be able to rely on the care of your bank for financial wellness, which requires a bank's good behaviour, above what is required by law and regulation. A bank that fails to comply with certain rules can easily pay the fine and establish new procedures to prevent the same thing from happening again. However, a dissatisfied client is unlikely to return and may take many others with them when issues arise.

"For the purpose of banking services, trust encompasses good behaviour of the bank above what is required by law or regulation", repeated one of the speakers at my conference, inspired by the paper "Virtuous Banking Placing Ethos and Purpose at the Heart of finance" (Llewellyn et al., 2014). That is, for me, what a new ethos is about. Ethics is good; it is great, but it is not enough.

Banks must be well-regulated and supervised, and diverse business models will keep the industry sharp and competitive. In that framework, one could argue that we no longer need ethos anymore. Yet, I argue, and many with me, that we need it more than ever. Today, we see that the pendulum of regulation is pivoting in the wrong direction, limiting some banks too much in taking up their role as a financial life partner. As such, sophisticated financial services regulations, not well understood by most people, or not understood at all, can even damage the trust these people have in their bank.

6.7 Conclusion

It is clear that in the complex society we live in, we need rules to understand the game we play. Banks have a tremendous responsibility, making them one of the heaviest-regulated industries for valid reasons. However, we should not forget that these limitations of freedom also come at a societal cost.

In every game, there are rules, and they determine your options to win the game. The choice you make to fulfil your goal is what makes you a powerful and respected player or not. That is equally true for banks. Regulations create a fair game for everyone willing to apply the rules. Those regulations are important, as they determine what can be taken for granted, so as a stakeholder, you can look at the other elements that determine your perception of an organisation. As a retail client, you don't need to worry about losing your €50.000 deposits if a bank goes bankrupt because of the deposit insurance scheme, but you may worry about the ease of accessing a helpdesk, for example, or about how those €50.000 are being activated by your bank.

Regulation will not set a bank's ethos; it is the bank's interpretation of a regulation that will make that bank successful and respected, that is, by doing good because they believe in it. In this context, I remember a quote from someone saying: "All bank employees must possess a sense of ownership by taking collective responsibility. Let's not forget that professional standards and Code of Conduct have been in place for decades. But this did not prevent the wrongdoing. So maybe the new management approach should focus on instilling a strong ethos, rather than just adhering to empty regulation and promoting decisions and behaviours guided by values that go beyond a company's legal requirements".

Ethos stands for the soul, the DNA of an organisation and having those aligned with what the outside world expects from you will make you stand out. That will drive your resilience and make you a trusted organisation. Putting ethos and purpose at the heart of everyday's actions is a way of taking destiny back in your own hands instead of having it imposed by policymakers and regulators. What I learned from the many talks on this topic is that the advent of sustainable finance and the evolution to a fully digital economy are giving banks the momentum to reshape their narrative and the image of their profession.

7

The Holy Trinity in Customer Experience

7.1 Introduction

Once upon a time, smart people found a way to convert the world's complexities into a binary world of 0s and 1s. 0s and 1s lie at the core of digital technology and have sparked a transformative revolution, a revolution that fundamentally reshaped how we communicate, conduct transactions, do business, work, travel, and manage banking operations. It has streamlined and revolutionised nearly every aspect of modern life.

It was never easier, never faster, to get whatever.

Creative minds searched for the best possible customer and user experience, and maybe even more for the highest convenience and efficiency. Layer over layer, development after development, created an evolutionary road to banking as we know it today. An internet and a mobile revolution later, banks have never been more omnipresent. You can check your balance or receive a payment at any moment of the day. You can set aside your money on a deposit account no matter when, and with some luck, you have a (virtual) assistant at your disposal to answer your questions, whether these questions are related to your financials or not.

Thanks to the race for the best-embedded finance services, your bank may show up for you, but only when you need them. In an ideal world, the customer doesn't need to contact their bank; they will join you on your journey whenever you need them. We can no longer ignore the omnipresence of financial services in our lives. For many banks, it is still with passive notifications, but a few have a much better understanding of the client and a much smarter system to contact clients at their convenience.

R. Coeckelbergs, *A New Ethos in Banking*,
https://doi.org/10.1007/978-3-031-78234-3_7

Most people use and actually enjoy using digital services to communicate with their bank. Pure transactional services no longer require a person in the loop in most cases. Reducing the branch density seemed a logical next step to free up expenses for more digital developments. It is a difficult balance to walk. Banks need to cut costs to invest in digital, so they reduced service levels for improved future service levels. After all, on the one hand, customers were not ready yet to see their branches disappear. On the other hand, they couldn't fully benefit from digital developments since they were still work-in-progress or because they didn't have the tools or digital knowledge to use it in an intelligent way.

Where the industry saw this as a logical evolution, many consumers felt they were being ignored. The road to digital banking was often framed as a cost-savings exercise instead of a customer service process. Even worse, the cost savings were often effectively the main reason to pivot to digital banking. The paradox of digital banking is that it was never more omnipresent and yet never more distant from some of its clients. Efficiency above effectiveness.

A C-level banker once told me, "Convenience is a dissatisfier". Efficiency has made banking services 80%–90% more convenient, but this hasn't always yielded the intended results. The industry has commoditised banking and built it so people take for granted that the tools work as designed and only show emotions when things go wrong. Banks have integrated fantastic user experience (UX) elements, but due to operating in silos, they rarely thought things through as some niche fintech players could. Consequently, we've seen many digital banking developments that didn't create moments of happiness. Customers often experience more stress from being unable to find their way, being overwhelmed by too many options, and feeling helpless in a world of endless opportunities.

Digital banking presents an incredible opportunity. Banks that prioritise convenience and excel in user and customer experience across their entire value chain have successfully embraced meaningful transformation. However, these banks are only a small minority. Fortunately, new developments on the horizon can assist the majority of banks in rethinking their transformation programme and integrating ethics and company values in a meaningful way.

In this chapter, we will try to find an alternative path. Let's call it digital transformation 2.0. For me, that is a digital transformation that does not just consider efficiency as the highest possible good but much more, one where effectiveness is perhaps the keyword, and empathy is being reconsidered. It is a transformation that many organisations are already applying today, some more than others.

We will look at how ethos—pathos—and logos can effectively become the holy trinity in customer experience, as Stefaan Roegiers, Head of Distribution at Banque Internationale à Luxembourg, shared with me: "You have to make sure that you offer a product that is correct, and you need to ensure that you can use data models that you can explain understandably to your customers. If we cannot clarify matters, then we will lose their trust. We will never be 100% digital, but customers may only come to us when they consider that we bring added value to them. That means finding the right balance between human, digital or phygital interaction. So, for me, the essence is that we need to find the right balance between ethos, pathos and logos" (Coeckelbergs, The Holy Trinity in customer centricity, 2024).

7.2 Generative AI: The Catalyst for Next-Gen Banking Innovation

We have witnessed the continuous hype surrounding various technologies. Some have had a profound impact on how the industry operates and conducts transactions, while others have had much less of an effect. Certain technologies, such as crypto, create parallel circuits, while others, like more standardised blockchain technology, alter the industry's processes.

I'm not a big fan of hype. I try to focus on the trends that outgrow the discussion, and I expect one of those to be Artificial Intelligence (AI) and, more specifically, the recent developments in this domain. Let me start by saying that AI is not new. It has existed for close to 70 years now. AI comes in many shapes and forms and has been shaping the banking industry for decades, one way or another. Robotics Process Automation (RPA) and hyper-automation are obvious examples of where a rules-based robot will take over some of the mundane manual labour or automate entire processes.

We see AI in robotics and expert systems; there is the subset of natural language processing and computer vision; there is machine learning and deep learning. Some of these are better developed than others. Banks have been testing, developing, and implementing many of these over the past decades. Most of the time, the focus of the business case was on making existing processes more efficient and lowering costs. It rarely addressed the core aspects of banking, and customer demand was seldom considered. Experts and decision-makers would assess a new digital tool or technology and integrate it into their existing processes, typically resulting in lower processing costs, increased speed, and more data without fundamentally changing the process itself.

Recently, we witnessed a spike in the development of machine learning and deep learning. ChatGPT took the world by storm, and it will change it forever. ChatGPT was the start of a new revolution that fits under Generative AI (GenAI) and focuses on content creation today.

I am convinced that in the very near future, GenAI has the potential to rethink banking. It is "a technology that can create content, such as text, images, music, and other media, like a human would. It uses deep learning models and large amounts of data to learn how to produce outputs that mimic real-world examples. This subdomain of AI is distinguished by its ability to generate new data points (content) from the patterns it has learned in the training data". It has the capabilities, if not today, tomorrow, to become everyone's personal digital assistant. That also explains the key use cases today which bankers shared with us: operational efficiency, augmentation of employee capabilities, improved customer experience, and risk management and fraud (Vorster, Generative AI in Benelux banking—Opportunities, challenges and future outlook, 2024).

Since GenAI set the strategic agendas for boards across the world, we have entered a rat race for new innovations. We come from an era where software development used to be 100% instruction-based. Developers set the rules to apply, and people understood what to expect from the interpretation of data. Today, this is changing. We have started to see the emergence of machines that can suggest how to build other machines. There are now systems that occasionally become more intelligent than humans and can interpret what they believe is important for their users. You give them a history, a prompt, and an enormous set of data, and the system will provide you with an output that is not rules-based but by learning its own rules. That output can be a concluding summary or translation, but it can also be an analysis of a set of data or even coding for developers in the support of their mainframe or an entire process flow, including output for a marketing campaign, to name a few. Mind you: they may occasionally be smarter today, but they will be smarter regularly tomorrow.

The output of machine predictions is presented as an answer, but it's important to remember that these predictions are not certain and should be seen as shades of grey rather than black or white. Even the most advanced tools only provide varying degrees of confidence in their output. The Global Financial Crisis has shown the consequences of blindly following such predictions without critical thinking or safeguards in place.

To properly evaluate and address the output of these systems, it is essential to incorporate risk management skills throughout the organisation.

Maintaining continuous awareness and attentiveness is necessary for making well-informed decisions based on the feedback from predictive tools.

As Andrew Vorster explained, "The genie is out of the bottle, and the race is on to integrate GenAI across financial services. The concept and technology may be all hype and headlines right now, but they will soon become "business as usual" probably much faster than you may think" (Vorster, Generative AI in Benelux banking—Opportunities, challenges and future outlook, 2024).

GenAI will transform how we look at digital transformation and require a different talent pool. As much as these systems are rational, they obey their users. Human bias will come back in the output. In a male-dominated industry, that means that systems risk reasoning like men. That puts even more emphasis on diversity to ensure an inclusive financial industry. But it takes more.

"Technologists think in a data-driven, mathematical, neutral manner", Joris Krijger, AI and Ethics Officer at De Volksbank, once said, "But algorithms are designed to distinguish; otherwise, they don't work. Moral considerations are inherent to their use. You cannot leave those to the developers of such a model".

None of that is possible without a solid and widely supported governance model in AI development, one that gives explicit responsibilities to everyone involved in the value chain. The same model must also explain the procedures in case of unexpected output. Joris explained that if you want to correct bias in the model, predictability goes down. So, he raised the question for everyone involved in AI developments: How much risk are you willing to accept to correct that inequality? These questions must be part of that governance model (Olthoorn, 2024).

Technical people think technical, but this new era requires more diversity to tackle these new challenges. Organisations will need to change the composition of the digital transformation teams. Today's digital team will have digitally trained talent, but it equally requires philosophers, and why not anthropologists? Today's digital teams need to understand marketing techniques, the opportunities of those techniques, and their limitations. They must master user experience and customer experience, as well as software development. On top of that, the team needs people with a more holistic lens on things, people who master the soft skills and know how to weigh ethical considerations, because technically everything is possible. In a future of endless possibilities and a democratisation of development costs, people must master how to build AND when to stop. We are living in extraordinary times where technology will bring new toolkits to deal with the societal challenges we face, and banks have new responsibilities to take. The people will set direction, make the trade-offs, and guide developments in line with a bank's ethos.

All of that will create but also require more creativity. We have tools to support our clients, as we have tools to support us to support our clients. Peter Hinssen described how the power of AI will make a big shift in the quality of work we produce in supporting the way we support our clients. Imagine we map the quality of service into bad, mediocre, and brilliant and that today, most of our services are mediocre, but we still have a decent amount of awful, and yes, also brilliant service levels. With digital assistants that guide us based on the history we feed them, much of the work will be averaged out. Awful service will luckily decrease thanks to that, but unfortunately, brilliant services risk being reduced since they deviate from the standard. It shows once again how important the interpretation of virtually generated support is. It is great to have a virtual assistant to make our lives easier, but that should not come at the cost of excellent human services (Hinssen, 2024).

Boston Consulting Group conducted a study in 2023 on the performance impact of using GPT-4 in the work environment and found that having a virtual assistant in a creative product innovation department enhanced performance for the bottom performers by 43% and for the top performers by 17%. However, for business problem-solving, both bottom performers and top performers saw their performance go down, respectively, 26% and 17%. The virtual assistant may suggest convincing but misleading answers based on historical data rather than accurate solutions for problem-solving.

GenAI's opportunities are endless, but they come with challenges. We have never had more information to consume, which is what GenAI is built on. We have never worried more about fake news, yet the new developments demand banks to be more critical than ever. That is equally important for the efficiency gains they wish to realise, and for the augmentation of employee capabilities they wish to create.

Klarna announced its AI assistant powered by OpenAI in February 2024. In one month, the assistant had 2,3 million conversations and was doing an equivalent of 700 full-time agents. This service was said to be more accurate in problem-solving, leading to a 25% drop in repeat inquiries, and better customer experience. Klarna expected this assistant to improve profits by $40 million for the entire year. "This AI breakthrough in customer interaction means superior experiences for our customers at better prices, more interesting challenges for our employees, and better returns for our investors", said Sebastian Siemiatkowski, co-founder and CEO of Klarna (Klarna, 2024).

The potential is much larger. We have not seen the full potential of GenAI yet. I saw someone summarising the current early stage saying we are in the "fart and beer app phase of GenAI". Like the Whip app, it was fun but just a gimmick. An executive at one of the round tables we organised predicted that it would take another ten years before we could fully grasp the opportunities that GenAI created for humankind (and banking).

What we have seen so far has mostly been efficiency gains but not real transformation. GenAI can teach new talent in banks to work on mainframe and COBOL. Just teach the machine how to programme and banks can keep using the legacy core, but GenAI also has the potential to change the core, the heart of banking. I wouldn't be writing a book if I knew what that would look like, but for the first time, we have technology with the potential to make itself more human (Candelon et al., 2023).

Can we build empathetic services through digital? Some will say you can, while others will say that you need humans in the loop. Several studies have shown that AI can make people feel more heard, and isn't that what empathy is about? Computers can detect signs that people may not see, hear, or feel and anticipate that much faster. A study in the US revealed that ChatGPT 4 answers to medical questions of patients were 3,6 times more high-quality compared to human answers and 9,8 times more empathetic. The lesson learned was that having a virtual assistant to support physicians would help them answer many more questions with a higher level of quality, having the professional review to avoid hallucinations or erroneous feedback from the virtual physician (Advisory Board, 2023). If that is true in health, why wouldn't it be true in financial health?

KBC Bank developed a virtual assistant, Kate based on LLM and found that people find it easier to talk to a robot on certain topics than when they talk to a person. Katrien Dewijngaert, General Manager Strategy and Transformation of KBC Insurance, explained to me in 2022: "What the data of the conversations and the aggregate data of Kate showed me is that talking to a bot lowers a lot of thresholds for customers. That was quite revealing to me, is that a bot knowing that it's not a person, not somebody who can judge you". KBC Group learned much more from its clients because they received questions that people would never ask another person, being afraid of the reaction. In the other direction, the bank now has a virtual assistant that reaches out to customers asking if they need help with an insurance claim after a storm, for example. It brings the bank, or the insurer, closer to the customer at the moment they need it (Coeckelbergs, The adventures of Toddler Kate, KBC's virtual assistant, 2022).

In 2024, KBC Group's clients raised more questions to the virtual assistant Kate than they asked employees, saving 200 full-time equivalents, on top of more effective customer service. "Where a human being can support 50 clients with damage on their house after a storm, Kate can do 1.000 in a few seconds and it can work 24/7" (Kanaal Z, 2024).

The essence of growing effectiveness in digital transformation lies in using digital tools to improve services in a way that makes people, including customers, employees, and other stakeholders, feel heard and well-served. This goes beyond just preaching organisational values. Embracing your ethics and values for a meaningful next wave in transformation will set your developments apart from your competition.

7.3 Logos in Customer Experience

The potential offered by today's technology is limitless, and all banks are exploring it in various ways. Progress may not be as fast as some would like, but the stakes are high, the resources are limited, and banks cannot hide that they are a risk management businesses that need to be careful with new technologies.

Banks have translated their business into the binary world of 0s and 1s, all nicely logical and rational. They rationalised processes and refurbished their business to be more efficient and convenient for both customers and employees. By going digital, banks could seamlessly trace every relevant change in the (financial) life of clients, from the number of contacts with a bank to the assets under management, the location of transactions, and so much more. Banking was never easier, contact with a bank became easier, and applying for new products and services became easier, at least for the cases that met the 80/20 rule. The 20% of exceptions became much harder. Deviations were at the bottom of the priority list.

Data became the new gold, and relevant insights became the Holy Grail. As with the Holy Grail, the interpretation of "relevant insights" is open for discussion. Talking about how to find the insight that matters was, and still is, much easier than actually finding it. The amount of data was endless but not always as useful as some expected. Personalisation and hyper-personalisation were terms abused for implementing sales-driven notifications instead of sentiment-driven customer support. Technologies like digital identity, Artificial Intelligence (AI), and blockchain can help banks fight AML and terrorist financing. They can also help banks score the creditworthiness of their clients and digitally onboard new clients. Applying AI in a bank has tremendous opportunities, but every minor mistake is penalised.

A recent study showed that banks are among the least trusted industries with AI; people don't trust them with their data. Call it an emotionally driven response to people's fear that financial institutions could abuse insights into their data despite the regulatory limitations and the excessive data abuse by other industries (like social media). Banks sit on a pile of data, and AI can translate that into valuable insights for the bank and the customer. To overcome the misconception of banks being unreliable with AI, it will be fundamental to build AI models and concepts ethically and in full transparency with the customer.

It is crucial to conduct thorough risk assessments of new technologies and understand the potential reputational damage if things go wrong. Banks cannot simply release new tools without fully grasping their impact. Launching a chatbot trained by large language models requires additional rules, control mechanisms, and an ethical framework to prevent hallucinations, incorrect answers, and discrimination. In the trusted environment in which banks operate, it is essential that the information provided is accurate to ensure that clients can trust the services provided.

Today, banks can no longer afford to set ethics aside in developing their new digital platforms because machines started thinking with us, but also like us. Their thinking is aligned with how people think and data provided to train them, including unconscious biases and misconceptions. To minimise unintended negative outcomes, banks need to establish an ethical infrastructure. Joris Krijger, an AI Ethicist, suggested that this infrastructure should extend beyond simply providing a framework for developers. It should include guidelines stating that models must be fair, explainable, and accountable. The ethical infrastructure must encompass policy, governance, tools, training, and assessment to ensure that critical thinking is not just a document but is clearly embedded in the DNA of the organisation and also in the role of employees.

Because GenAI is such a scale business, most banks will not build these solutions alone. They will depend on others; this brings additional challenges if they wish to counter the preconception that most people have about banks and the potential of data insights. Banks will need to talk and agree with suppliers on how to find a common ethical infrastructure and interpretation of the possible, one that makes sense and one that banks can be accountable for (Vorster, The AI Stratego—Defining the rules for banking, 2024). They will need to define the rules of what order and reason will look like in that future.

"The Gartner hype cycle that defines the maturity of a technology can be a benchmark. In my opinion, a bank always needs to step in when the maturity slope has passed the hype wave. That is the sweet spot where risks and advantages for customers are moving to an acceptable balance. So, banks should not be the digital or the tech leaders. That's not our role, but we should make sure that when we use new technology, we give our customers access to it in a secure manner", explained Stefaan Roegiers in his interview (Coeckelbergs, The Holy Trinity in customer centricity, 2024). Showing the right intent and acting upon it is the only way to persuade customers that positive change is achievable.

Banks will need to learn how to build more complexity but in an explainable way. The current wave of innovation through Generative AI (GenAI) is a tremendous opportunity for the industry. But it comes with new responsibilities. 80% of data in a bank today are unstructured. GenAI allows to mine these gold mines, but high data quality is critical to get the gold out of there.

The industry's potential with GenAI is boundless when coupled with a robust data framework and effective governance. We have already witnessed the impact of robotics process automation and hyper-automation, significantly reducing mundane tasks for employees. Today, with GenAI, the impact is expected to be even greater, as virtual assistants can now provide support in more innovative ways. They can assist with tasks such as drafting emails, displaying pertinent information on screens during client meetings, and providing real-time sentiment analysis during calls. GenAI enables clients to receive 24/7 support on a larger scale, offering highly personalised advice and assistance rather than just location-based weather notifications for insurance.

This is just the start of a new wave of developments. Because the output leaves the machine so client-like and human-like, the biggest risk is that people forget that it comes from a machine. Logos has never been more important for banks because the answer can be life-changing, and that means that it needs to be explainable at some point, which requires strong governance and a level of data integrity we never expected before. Banks have mostly focused on high-quality data of what was structured, the 20% and today, the focus must be on the remaining 80%.

7.4 Pathos in Customer Experience

Customer experience used to be straightforward. People who had access to banking services knew what to expect. It was a very human service involving people talking to people and making deals. Banking is a human business, and while this hasn't changed, the way banks approach offering human services has. The interpretation of what human services involve has also changed. If you were to ask two people to define "human service", you might receive two different answers. This is why "Pathos" and empathy are fundamental for banks in offering services in a way that their clients appreciate.

People above 40 grew up in a different bank context. They knew their branch advisor, and they trusted them. They knew the branch advisor and the branch advisor knew the customer. That meant people took the advice for granted; there was a connection, a human connection. People under 40 may appreciate using an app for their banking transactions, as long as it is made easy for customers. It's important to have the option to speak to a human if they encounter a problem. They may not need much human interaction beyond that, but not everyone feels the same way. How can a bank establish a trustworthy brand image when it needs to serve all clients efficiently, and the quality of both digital and human experiences can make or break the bank's reputation and overall approach to digital banking?

There's no turning back from digital transformation. The future is digital, and technology is advancing to a point where we can consider a new phase in this transformation. However, not everyone is ready for today's technology, at least not at the pace of today, and banks must consider this.

Ghela Boskovich once said: "Emphatic services drive business". This is a call to action for bringing pathos back into the equation: How can we turn 0s and 1s into services that embrace the needs and perhaps even the dreams of the ones being served?

The opportunities presented by technology today can help banks make their digital services more empathetic and human. Banks that achieve this have gold in their hands. Not only will this lead to increased income, but it will also foster stronger relationships, resulting in more loyalty and less customer turnover.

Technology should not deprive us of our humanity, as it has done in the past, Ghela explained: "The human touch in developing new technological concepts allows us to design services that mean something but also deliver something. The value of those deliveries is processed through emotions, not just pure economics. So our perception of value is also based on the perception of us being understood. That perception is what empathy actually is".

Many industries, including the tech industry, aimed to digitise their existing processes in the most customer-friendly way possible, often while stimulating cross-sell opportunities, but this approach was still a one-size-fits-all solution. And most of the time, they constructed piece by piece instead of reconsidering the entire platform (which is nearly impossible to do anyhow). Unfortunately, this is not the right way to create truly user-friendly services.

Most banks cater to anyone with money, individuals, and corporate customers. Each of these customers has unique desires, wishes, and expectations. In the past, branch managers used to understand their customers personally, based on years of experience, shared stories, and lots of coffee. They tailored their approach and personalised it, to each customer based on their individual needs. However, with the rise of technology, and also heavy procedural regulation, most banks lost the human touch that used to be a hallmark of banking. The key use cases were generalised and built to serve all similarly, simply applying the 80/20 rule. Because scale means a lower cost-to-serve. Also, beyond technology, banking evolved into an expert industry that demands so much from people that it has become impossible each advisor to know everything.

Technology used to require technical skills, often only that. The business side of the bank would explain what they needed, and IT would translate it into code. In the past, banks often forgot about the psychological, sociological, and anthropological aspects and skills. At best, banks thought about digital banking like bankers or technologists, people sitting in the head office and putting themselves in the place of their customers, different people than themselves.

"Empathy is a component of trust. Empathy is respect. Empathy is a component of value exchange", is what Ghela said. Value exchange is what banking is about, so empathy is an integral part of that business, and it is time to get empathy back into digital banking developments. Bringing empathy back into the equation starts with one key question: how will you become a mirror of society, or at least for the customers you serve? How will we find out how they think, what their challenges are, and how they relate to money? What problems can you help them with, and how do you solve those problems in the right way?

Don't make the mistake I used to make by assuming you know because you don't. Those assumptions start with a lot of biases, and a bias is like a hall of mirrors, where walls are lined with mirrors that distort reflections in various ways but never truly capture the other side of the mirror.

Becoming a proper mirror of society takes an open mind and solid research. Beyond the traditional approaches like interviews, focus groups, user testing, and surveys, banks can still learn a lot from, for example, ethnography, a method of studying people in the places where they live or where the action takes place. Leaving aside the vacuum and framing of traditional research tools may provide much more honest and less biased feedback for banks to build on. Ethnography will add qualitative research on top of the quantitative research. Observations can help to cross-check internal findings and add an extra layer to empathetic feedback that is hard to measure in multiple-choice surveys or focus groups in separate rooms (Broløs & Taylor, Customer-Centric Innovation in Finance: Leveraging human insights to drive product innovation in the digital age, 2024).

Logos is all fine in building the best possible bank, rationally speaking, but when empathy is forgotten, the business makes no sense.

GenAI revolutionises personalised services by providing unique answers to every question and tailoring approaches based on the sentiment expressed during conversations. It can address frustrations and ensure customer satisfaction without compromising profitability. Over time, GenAI can deliver increasingly customised support when evaluating creditworthiness or optimising investment portfolios. In the future, loyalty will play a significant role, and hyper-personalised service can strengthen customer loyalty and profitability. Empathic services as a revenue driver.

So let us embrace technology and figure out how to serve clients, and perhaps even the community and society, in the way they wish to be served, whether that be on the basic banking needs, more sophisticated financial well-being tools, sustainable finance initiatives, or digital inclusion. But first and foremost, let's start listening with an open mind to the customer and translate that into business opportunities to build lasting, long-term, and loyal relationships.

7.5 Ethos in Customer Experience

Now we know what to consider to build a bank's arguments to strengthen its logos and its pathos, it is time to consider ethos. Customer experience is not a science, but it gets close to it. It requires a predetermined set of principles. The same is true for renewing the connection with the customer, but as important as both are, the magic is in translating those insights into the third element: the bank's moral framework, its identity, and values, its DNA, or as we name it in this book: ethos.

To define a new ethos in banking by embracing values and ethics for a meaningful transformation, banks will need to translate their values into their day-to-day business. That happens increasingly online, but not always. That is why I talk about customer experience and not digital banking.

For example, if a bank values accessibility, inclusion, and engagement, it must find solutions to help customers who struggle with technology, whether these solutions are part of the digital package or not. This group includes people with disabilities who need specially designed user experience, who may need screen reader optimisation, higher contrast modes, voice banking, or accessible rich internet applications.

This group also includes those who are not familiar with digital processes or don't have the tools to access them. Financial exclusion is a concern, for sure, but digital exclusion is often even worse. People who are digitally excluded unfortunately have one foot outside of the social cohesion that most of us experience. Banks that aim to build a strong ethos based on digital customer experience must find ways to support the fight against digital exclusion. There are numerous examples of such initiatives. Some banks seek support from supermarkets or postal offices, while others collaborate with press shops or libraries. Sector federations may organise roadshows and training days to assist people in learning how to bank digitally. Some even have mobile teams that meet clients in their homes to guide them in the right direction.

By supporting initiatives against digital exclusion and financial exclusion, banks are taking a more thoughtful approach to ensure that more people can benefit from the significant investments they have made in the past. This shows that they are taking a more thoughtful approach to ensure that more people can benefit from the significant investments they have made in the past.

Next to that, we may all find ourselves in the situation where the obvious does not seem to work as designed. When things don't work as intended, who do you turn to? Being put on a waiting list in a contact centre is not the right approach, and online chats can be even worse, as they require customers to stay glued to their screens in search of an answer.

Technology can help bridge these gaps. Tools like GenAI could assist in identifying and solving these issues. The technology can help find the gaps in the process to avoid people getting into a process that doesn't work as designed. Additionally, it can help find the best way to support the customer in a human-friendly way. It can help to understand the problem, and with a bit of sentiment analysis, the technology could even measure the sense of urgency to evaluate which channels fit best to support the customer in the best possible way.

Effectiveness by design, using technology to build brand authority and digital supremacy. Or am I using too many superlatives now to describe the value for banks to properly use GenAI?

Euroclear is not the typical retail or wholesale bank, but the message of Kevin Johnson, Head of Innovation Competence Center at Euroclear on GenAI is equally valid in retail or wholesale banking. He said: "Our current view of Generative AI is this is an augmentation to human capability, not a replacement of it". Whether it's about translating legal documents into plain language, like at Euroclear, assisting the service desk in providing the right answers to customer queries, or aiding in investment discussions, technology is enhancing the work of humans. This is about efficiency, sure, as employees will spend less time on similar work, but beyond that, it should also help customers faster, it should give the customers a better, more contextual investment advice, or help better understand legal implications of new regulatory requirements (Vorster, Generative AI in Benelux banking—Opportunities, challenges and future outlook, 2024).

In that respect, GenAI is making banks not just more efficient but also more effective. As a client of KBC Bank, I have experienced how these tools can help guide not-so-digitally savvy people in the right direction. Since 2020, they have had the virtual assistant Kate "to help you with all your questions". If I cannot find something, Kate can help, and if she can't, she'll put me in contact with a human advisor.

I believe technology has reached a point where we can rethink its potential. Over the past 20–50 years, technology has been more of a cost saver, creating more distance between banks and their clients than closing it. Branch advisors were sometimes spending more time educating people how to bank digitally. Today digital tools can educate people how to bank digitally, or even make it so convenient that training is no longer required.

Today's most sophisticated bank app allows you to manage your entire life, from money affairs to public transport, house renovations, lending products, payments, lifestyle, and loyalty to sometimes pure entertainment. Everything is there and more, to the extent that customers sometimes get lost in the haystack of functionalities. Does this 'ecosystem' of digital life services still align with what people expect from their bank? Does it contribute to a bank's ethos, or is it too far from the perceived purpose of a bank?

And is this a relevant question to ask an industry, or should it be answered individually, where some banks can rightfully answer the question differently than other, more conservative brands? I believe it is a highly individual set of questions that every bank should answer for themself, just like how they want to frame their ethos. What level of digitalisation fits the DNA of the bank and how will you translate that into your services? How do you augment your values in the digital services you offer? How will you balance digital and physical, which I don't expect to go away anytime soon?

"We need to ensure that we keep our customer's trust in the real world and in the digital world. As such, banks should aim at being customer's trusted partners in the Phygital world. Phygital? Yes, our customers want to be serviced in both digital and physical reality in a balanced, efficient manner.

Banks need to ensure to keep this trust in both realities", shared Stefaan Roegiers with me (Coeckelbergs, The Holy Trinity in customer centricity, 2024).

Being a trusted financial advisor comes with responsibilities and expectations. That includes a compelling app that allows anyone to manage their finances whenever, wherever, without any concessions. Every bank has values and promises to its customers, and these need to be translated into the digital context to ensure you build a consistent brand value across all channels.

If sustainability is high on the agenda, ensure that it can be felt in the digital bank environment, and help your customers reduce their carbon footprint, stimulate them in responsible investments for greener living. Don't limit yourself to a carbon emission calculation based on daily banking transactions. Provide them with a platform to streamline the experience of using public transport. Help them calculate the emissions of the house and provide reliable suggestions to reduce them. Make sure that your investment strategy breathes the same aspirations. Take inspiration from Alipay by linking loyalty to planting trees instead of useless loyalty points. Help the customer to reduce its carbon footprint beyond the obvious.

If you are an innovative bank that "is there to realise people's dreams", make that clear in the digital environment. Help people to save for dreams. Savings goals can be a helpful tool, ideally starting with visualising the dream connected to a savings account. Add smart gamification elements to stimulate your clients to save for those dreams and offer valuable tips and tricks to save more. Build an experience around how you realise dreams. Gamification is a relevant approach, but I'd also argue to use "storyfication".

Building stories around the dream will make people more engaged and adding game elements may stimulate them to change their behaviour to do more. It will make you stand out from others, but even more: it will show you that you are serious about realising your dreams. That is what will grow your reputation, trust, and, with that, your ethos in digital experience.

Small businesses want to do business and hate most of their administration. Banks are uniquely positioned to unburden businesses in every possible way. Being the best SME banking services provider has become more challenging with the rise of embedded finance. It's much more interesting as well, by the way. Embedded finance brought banks closer to their customers. It offered some of those customers new revenue streams and cross-sell opportunities by connecting bank APIs to their platform, allowing them to sell financial services themselves or offer credit solutions in their sales processes. Banks that properly deploy embedded finance will be there the moment a business is being set up, like KBC Bank in Belgium, which connects with social secretaries to allow them to open a bank account the moment an individual applies for a company. In Belgium, it is common to take a company car, and with a bit of luck, KBC Bank will be embedded in the software of the car dealer for an instant lease approval at the moment of purchase.

More common is the links with accountancy software and payment solutions. We spoke to Barry O'Sullivan, Head of Banking & Payments Infrastructure at OpenPayd, who said: "With our API, customers can easily integrate with OpenPayd, selecting from an array of financial products. It might be that they want, say, multi-currency accounts, or they might want global payments, Faster Payments, or SEPA access. They can pick and choose which products they want to add and then obviously offer those out to their customers" (Coeckelbergs, Unveiling the future of finance—A comprehensive exploration of embedded finance perspectives, 2024). There are many more examples like that.

More universal banks face additional challenges in that whatever is provided through embedded finance APIs can also seamlessly be accessed in other channels as part of a design strategy that focuses on effectiveness rather than efficiency. Koen Adolfs, Head of API Portfolio, Future Customer & Payments Domain at ABN AMRO Bank, shared in that respect: "Embedded finance is not a stand-alone proposition. Propositions should work seamlessly together across their own channels, solutions and third-party integrations. This means many things must come together to offer a successful API product to the market".

This shows that morality, on the one hand, and a deep understanding of customer needs today and tomorrow will ultimately lead to a meaningful transformation that interprets technology as a means to an end rather than the ultimate goal. Building a trusted brand will make your business future-proof. Technology will, without a doubt, deliver the tools, but we have outgrown the period of digital transformation now.

Let's agree we are digital, and now it is time to build meaningful services that look beyond efficiency and consider a more effective view of developments.

7.6 Conclusion

"Ethos in banking is not going to hold back banks, but it's actually going to be a facilitator for banks. As long as we can prove to our customers that we are ethical, that we have the right intent, we will be able to earn their trust, and we will be able to do our jobs. But once we lose that trust, customers will turn to other solutions outside banks and finding these solutions is easy these days. So being ethical will become a differentiation factor for banks and a key element in the relationship between banks and customers", explained Stefaan Roegiers, back then Head of Business Transformation at Banque Internationale à Luxembourg.

Digital technology is an integral part of life. Although the book is not about digital transformation, it is impossible to ignore it when talking about embracing ethics and values for a meaningful transformation. With new possibilities through technological advancements like GenAI, we are at a new crossroads of reinvention.

I repeated a few times that convenience is a dissatisfier, and that may be one of the key considerations for stopping to look at digital transformation for more convenience. We have passed the period when efficiency stood at the core of digital transformation.

Banks shouldn't lose sight on cost reductions in transforming their business, but it shouldn't be at the core of transformation. By applying a more effective approach to changing the business and by focusing on the bank's identity, values, and morality even, technology can be a powerful set of tools to enforce that DNA in the transformation in the business, leading to a more sustainable business model.

Digital transformation should no longer be about adding more gimmicks and increased automation. It is time that banks approach their next steps in their transformation in a much more holistic approach. It should be an approach that looks at the long-term instead of the short-term cost reductions. By putting doing what is good at the heart of this, by putting the client at the centre of this transformation, the opportunities are not just cost reductions and cross-selling opportunities but will lead to long-term customer engagement.

By embracing ethics and values by redefining a bank's ethos for a meaningful transformation, banks can reconnect with their clients and society again, as they did in the past, long before every bank offered kind of the same services in an online environment. It is an opportunity to differentiate themselves from the rest, based on a deep understanding of what the organisation stands for and what that means to its clients.

GenAI is set to transform the banking industry by automating routine tasks, allowing employees to focus on more valuable activities. It will enhance both employee and customer experiences by analysing vast amounts of data and understanding customer emotions. However, success depends on proper training and an ethical governance framework. Banks will need additional technical talent and must prioritise efficient implementation to avoid standardising mediocre service levels. Emphasising the importance of customer experience, GenAI promises to make new or existing tools more accessible to banks, and allow for much more personalisation of the digital bank experience, on a bank and a customer level.

Money is as emotionally loaded for most people as it is a rational concept, and in customer experience, that is too often forgotten. So, let's deepen our understanding of the customer and get to know them much better and at a much bigger scale. Each person has their own relationship with money, and if banks want to build delightful customer propositions, those relationships need to be properly understood and considered in developing those propositions.

Empathy is a component of value exchange. An empathic company will build more long-lasting relationships that, over the long run, will generate a higher customer life cycle value. Today, more than in the past, technology allows us to get more from stronger customer relationships, making the customer feel understood and appreciated and with stronger ties to the corporate brand.

That is where ethos in customer experience comes in. Embracing ethics and values in a meaningful transformation means that banks weave their values into how they do business. Actions about corporate values, not words, will make banks more trusted again, building longer-term relations that are not purely based on the best rate but on a common understanding of the role a bank has in the life of people and society at large.

Trust will come with the delivery of services (ethos) that customers expect from you (pathos), in a way they expect it, with the right duty of care and expertise (logos). Does that mean banks should no longer build ever more convenient services? Of course not, but they must be cautious to ensure that everything works, not just as designed, but also as intended. Luckily GenAI can help with that as well, that way avoiding that we keep talking about "Convenience is a dissatisfier", but that convenience rises to become a unique selling proposition.

As we said at the beginning of this chapter: "You have to make sure that you offer a product that is correct, and you need to ensure that you are able to use data models that you can explain understandably to your customers. If we cannot clarify matters, then we will lose their trust".

8

The Biggest Opportunity Since the Industrial Revolution

8.1 Introduction

The Green Transition will affect us all, whether you like it or not. In fact, it already does. Our current way of living is no longer sustainable for the planet. Scientifically speaking, change is imperative and should not be underestimated. This may be the most significant transformation in human history, necessitating not just policy and mindset shifts but also substantial funding.

That explains why this Green Transition is often considered the biggest opportunity for the banking industry since the Industrial Revolution, an era in which banks (pre)financed most of the railroad investments and electrification of the world. Today, humankind is in an evolution of an even bigger magnitude and with a much bigger complexity.

On the one hand, we know today where we want to end up by 2050, but the path to that destination remains blurry. Technology continues to evolve, and we may discover new solutions to address the problem, but we do not know yet. Clearly, innovations will play a key role; however, for the benefit of the planet and future generations, we cannot afford to wait for new innovations to emerge to resolve the issue. We must act now.

The goal of the Paris Agreements was to keep global warming below 1.5 °C, which translates in a reduction of 37 gigatonnes from 2022 levels and achieving net-zero emissions in the energy sector by 2050. Under the current plans, the world will fall short of that, with a reduction of 16 gigatonnes instead, not even half of what is required. Global renewable energy share must increase from 16% in 2020 to 77% by 2050. That is massive, especially considering the long investment cycles in energy. More than the

R. Coeckelbergs, *A New Ethos in Banking*,
https://doi.org/10.1007/978-3-031-78234-3_8

investments in renewable energy, minerals will also be incredibly important because these assets are the baseline for many new innovations in the renewable energy and battery industry. Innovation will help the world to use energy smarter, through various smart metering and management systems, with the Internet of Things as the glue between them.

That means tough and unpopular investments are needed, which may lead to social turbulence. Housing will need to be modernised with a zero-carbon footprint in mind, involving massive upfront investments and a long-term business case. Implementing new pricing mechanisms and carbon taxes to shift consumption and decarbonise transportation will lead to limited choices and more control, but it is the only way to create the necessary shock to preserve our planet. This may contribute to growing inequality and a decline in social welfare.

Knowing that the Green Transition can only succeed if everyone joins the train of change makes it extra challenging, but also an opportunity for banks that embrace ethics and values for a meaningful transformation. What does that mean to banking and how can banks approach this challenge, limiting social turbulence as much as possible, witing their circle of control?

What is that circle of control and why have small banks a luxury to apply more aggressive strategies in this respect compared to the big players?

8.2 The Green Transition Versus the Industrial Revolution

The path to 2050 is long; it almost seems endless, but the timeframe is frighteningly short. The change needed has never seen before in human history, if we want to offer a future for the generations to come.

Something similar was said about the Industrial Revolution. The impact of steam should not be underestimated; it allowed factories to rise, trains to roll out, and boats to transform the industry as people knew it. The Industrial Revolution changed traditional careers, moving jobs from primarily agrarian to manufacturing, which led to rapid urban growth and also more pollution. The massive investments to achieve all this made banking more important and more developed. Social tension and labour abuse led to labour movements. Mass production changed consumption, and the telegraph transformed communication.

Needless to say, how much this affected humanity. So, to what extent can we compare the Green Transition with the Industrial Revolution for banking?

Why is the Green Transition so important for banks to demonstrate to society that they should be seen as enablers of change instead of value extractors?

Let's be clear: the Green Transition differs from the Industrial Revolution in several ways. During the Industrial Revolution, only a few people had the opportunity to invest in railroads and industrialisation. Consumers and employees were simply confronted with the change and changed accordingly. The financial industry made a fortune from a few large clients, but it was not a retail business. In contrast, the Green Transition requires every bank to participate and bring their clients along on this journey actively. Everyone has an active role in today's transition, with an immediate effect on day-to-day life.

On the other hand, 200 years ago, the Industrial Revolution was a green field project. The investment needs didn't replace an existing infrastructure. There was no legacy to break down and much less resistance against change to fight. The transition today fundamentally impacts everyone's life and is not always for the benefit of people's situation today. It is essentially a struggle today for survival tomorrow. That message is much tougher to convey. That makes the situation a tremendous opportunity for banks to show how important they are in that transition and prove that they do it with the best intentions, respecting ethics and taking up the responsibility they are granted to facilitate the Green Transition.

The transition requires massive funding, volumes that cannot be achieved spontaneously, but not purely through public finance either. The International Renewable Energy Agency estimated that a net-zero economy will require approximately 150 trillion USD globally towards 2050 to achieve the 1.5 °C target (International Renewable Energy Agency, 2023). That is impossible to do without the healthy financial intermediation provided by the financial services industry. The alternative of not reaching the target, which cannot be considered an alternative, just dealing with the cost incurred by the higher temperature levels if we continue living as we do today, is estimated to be 2,328 trillion USD (Climate Policy Initiative, 2023).

Sustainability is a general societal concern that touches us all. The warnings of global ecological crisis are not new. The first research studies showed this already in the 80s and got additional attention after the publication of a United Nations report from 1987 that became known as the "Brundtland Report", named after the chairwoman of the World Commission on Environment and Development" which published it. The report highlighted the global warming and its impact, it documented how it evolved from being "a side effect of industrial wealth" to a survival issue for many.

The problem was that society seemed to miss a certain sense of urgency, focusing much more on short-term improvement and ignoring the devastating long-term picture despite the call to action in the Brundtland Report: "Unless we are able to translate our words into a language that can reach the minds and hearts of people young and old, we shall not be able to undertake the extensive social changes needed to correct the course of development" (World Commission on Environment and Development, 1987).

To build tomorrow's sustainable future, banks, large and small, must demonstrate leadership. They need to sail the ship in the right direction. They have the tools and the responsibility to drive a societal change, a transformation to a greener world. The financial services industry was the oil that kept the wheels turning to a successful industrial revolution, and today, that same industry can be the wind of change again to activate both funding and responsibility to contribute to a better future.

That will require a mindset shift, one that begins with regulation. Banks can approach this defensively by simply complying with regulations, which is the most cautious way to handle the transition. Alternatively, they can proactively apply the laws to raise awareness about climate change among their clients, get them on board, and help them understand their impact on the path to achieving net zero. Instead of creating or selling complex products to make people unaware of how their money is being used, they could act transparently and make people feel responsible for how their money is utilised.

Remember what we said earlier: banks' actions today will influence how society looks like in 10 years. Banks have a big responsibility in shaping our society towards a more sustainable and green future. That isn't easy, and it triggers a lot of debate, both within the sector and the industry, in discussions with its stakeholders, like politics. Banks are taking these responsibilities seriously. Some conversations and presentations make me feel that banks have learned from the past and wish to do better going forward.

Now that global warming affects us all, the sense of urgency is changing, not just in banks but also on the street. People feel the impact on their lives, some more than others. The Green Transition is not just an ethical consideration but also an economic one. Banks and financial institutions like insurance companies have a particular self-interest in this societal problem. But what happens if specific industries suddenly face unexpected consequences of global warming or measures to reduce global warming?

Suppose you are a bank that invests heavily in agriculture, and suddenly, the areas that the bank invested in are no longer fit for farming, or the measures taken by governments to reduce carbon emissions are so restrictive that it no longer makes sense to keep farming.

A bank's real estate portfolio serves as a prime example. Banks help people achieve their dreams, and one common dream is to own a home. Physical risks associated with extreme weather can impact households, businesses, and their assets, affecting their likelihood of default. Additionally, there is a transition risk associated with the shift to a low—or no-carbon economy, leading to stranded assets that lose value over time due to decreased demand.

Banks that grasp the Green Transition to rewrite their narrative will need to consider a variety of initiatives. They obviously need clear strategies on the investment side. Which projects are they willing to fund or finance, and at which conditions? What will be the long-term impact of these projects on the path to net zero? What measures can banks take to promote more sustainable initiatives, like better financing conditions or higher funding?

The moral compass is more complex than ever because giving priority to the Green Transition may harm the efforts that banks wish to make in the "S" part of ESG, which refers to social measures. If a global bank suddenly stops investments in mining, it can potentially affect the employment rate of entire regions. There is not one right or wrong answer on how to deal with this issue, but a bank must make the choice consciously and with respect for its values and priorities.

The financial industry has long addressed global warming as part of its ESG strategy. Recently, it has gained more attention. Banks and investment funds employ strategies such as negative screening, avoiding investments in certain industries, and positive screening, actively investing in companies that meet specific criteria. Additionally, the engagement strategy involves taking a stake in a company to influence shareholder meetings and improve its behaviour (Cristina, 2022).

The transition strategy is another common term banks use to explain their efforts for a greener future. This strategy, often considered by the big banks, aims to leave a lasting impact without excluding by default. The only condition is that the beneficiary has a plan to support the path to net zero. In that respect, a bank may finance a fossil fuel company, but only if that company has a clear plan for an alternative future with low to net zero. As we will learn, this should be a plan that is evaluated regularly with consequences if specific targets are not met.

Some banks will need to consider even more stakeholders involved. Some countries and entire regions are highly dependent on fossil fuels, and the financial flows are crucial to sustain the economy. Balancing a healthier planet with local employment presents an equally significant challenge for certain global banks. A transition strategy should boost the local economy towards renewables, resulting in new unintended consequences. This places banks in a challenging position when crafting a narrative appealing to both progressive and traditional voices.

Moreover, banks have a variety of other options to assist their customers in reducing their carbon footprint. The most common ones involve encouraging clients with improved rates for renovations and electric cars, for instance. These measures are, by the way, beneficial not only for the client but also for the bank as part of their risk measurements in future-proofing the assets they invest in. But there are much more possibilities with the assets and network available to them.

Regulation is the minimum requirement and, therefore, is not worth discussing in further detail. When redefining their ethos for a better planet, banks will need to make conscious choices about going the extra mile, how they want to serve their clients and when to stop serving them. It must be a strategy that holds the bank, its management, and possibly its clients accountable for their actions and ensures that the money spent serves not only the clients' individual wealth but also the environment. More than that, it must provide the tools to ensure that the strategy is, more than words, becoming a way of doing business.

In short, redefining their ethos by focusing on the Green Transition requires banks to find a balance between logical and scientific facts and risk measurements (logos), an empathetic approach to onboarding clients in this global change (pathos), and effective action that results in a concrete change in the bank's culture and DNA, as well as in the community and the planet it serves (ethos).

8.3 How a 1-Euro Three-Sided Coin Makes a World of Difference

Vdk bank is a local, community-focused bank in Belgium. They were founded in 1926 by the Ghent-based workers association and a few local families with a very clear purpose: to improve financial inclusion and to stimulate more emancipation in an era with minimal labour rights, a time when social security was utopian, and many people faced financial stress. Men would receive their salary and head to the bar first before facing reality at home.

Leen Van den Neste, vdk bank's CEO, explained that history by saying: "We wanted the workers to save a little bit of money for their household.

Women were not working those days; it was a man providing for their households. We wanted to push them to save some money" (Coeckelbergs, Banking for good—A vision by vdk bank, 2021). Still today, vdk bank is a strong ambassador of value-based banking.

In 2023, they surveyed 1000 Belgian savers and investors and concluded that 7 out of 10 don't know what their bank does with their money (GABV, 2023). People who are unaware of how their bank uses their money probably don't realise the potential impact of their money. As a result, these individuals may be less receptive to messages from banks with a clear purpose, as they don't feel personally addressed. That conclusion was the start of a very successful awareness campaign.

With permission from both the National Bank of Belgium and the European Central Bank, vdk bank minted its own coins, with three sides instead of two. The objective of that campaign was to encourage Belgian citizens to think more deeply about how they can utilise their money to build a better future for themselves, their community, and the planet they live on. The coins served as a conversation starter with customers, prompting them to reflect on finances in a non-binary manner. This coin didn't just have two sides to determine which team would kick off a game. It didn't just represent win or loss, debit, or credit.

The three sides of the coin symbolise the value your money can represent for the customer, for the community, and for the planet they live on. Customers can expect a competitive interest rate from their bank, but there is more to consider, and that is how a bank activates the money on deposit accounts and investment products. vdk bank focuses on the local community in the provision of loans. They apply strict exclusion rules in their investment strategy now, although that was not always the case. They don't take the easiest route to success, but they are consistent and clear in what they say, which aligns with their actions.

Climate change is a real concern today, which means it is also a priority for the community and even more for the planet. So, although they invested in fossil fuels in the past, they explicitly don't today. They are very vocal about what they invest in. All that information can transparently be found on their website, both the methodology used and the companies they invest in.

"I believe that as a banker, you must invest the money customers save with you in ways that align with their preferences. This should be the new ethos of banking, regardless of the type of bank - whether it's a digital bank, one like vdk bank that offers both digital and personal services, a big bank, or a universal bank. It's crucial that you invest your customers' money according to their wishes", explained Leen when I spoke to her about the project.

The campaign received a massive uptake, both locally and internationally. The unique coins travelled the world. Chris Skinner received one, but also the Belgian Minister of Finance and Christine Lagarde, for example. It became a real conversation starter and an acquisition tool for new customers (Coeckelbergs, The triple-sided coin and the new ethos in banking, 2024; Coeckelbergs, Banking for good—A vision by vdk bank, 2021).

I believe the main reason for the success is that this small coin made the theory behind banking and money very visible and literally very tangible. By bringing the story this way, the concept of banking was much easier to explain to the average man on the street as well. Logos in the language of the customer (pathos) led to a convincing story based on a consistent message and behaviour (ethos).

The three-sided coin brought banking and sustainable investments much closer to vdk bank's clients by saying that they want to use savings to offer a competitive return to clients while also intermediating those savings in local investment projects that respect the planet and/or contribution to the community cohesion (as an answer to the S(ocial) element in ESG). That makes their story much more concrete compelling and convincing. It may not have redefined the ethos of vdk bank, but it surely made their ethos much stronger and stood as a great example for other banks to do so.

8.4 HSBC's Green Transition

I heard in several interviews that small banks and cooperative models have an easier path to a new ethos compared to more dominant and international banks. Guy Schellinck, CEO of Beobank, explained his context, for example, by saying: "Our bank, Beobank, operates under the cooperative spirit, even though we are not a cooperative bank ourselves. Our shares are entirely owned by Crédit Mutuel, which is a cooperative bank. This enables us to align our mindset and strategic objectives with those of the group. The cooperative banking model offers several benefits. For instance, we do not have any shareholders, except for Crédit Mutuel. We have fewer short-term commitments. This allows us to focus more on long-term projects and investments that benefit society, instead of prioritising short-term profits. Therefore, we are committed to using our resources to make a positive impact on our customers and the communities we serve".

Corin Millais, The of Socially Responsible Banking at Teachers Mutual Bank in Australia, went one step further: "Our members - our customers - are our shareholders and we don't serve two masters but one. We can think longer term, and no daily share price obsession is a huge gift. Who I answer to is the member – that's our primacy. Everything we do has to be for the member first, in the same way as a listed bank, which is all about the shareholder first. Their message of being customer-centric is questionable - it's all about returns. There's no dividend; it is retained. We're not sucking money out of the business. And no surprises then that people want us to look after and respect the community and the environment". He was also the first to say we also need "the greedy banks" to facilitate the big green deals and closed our talk by saying: "The planetary balance sheet is in the red and has to be reconciled. And we're going to need banks to do that – in fact, I think, and it's been said a few times, that the finance sector will save the world in the end. That's the ethos we need – but you've got to build it with your heart and as you say - pathos".

So can we conclude that redefining the new ethos is left to the small banks and making the rational big profits with the Green Transition by the big banks? I don't think so. Ethos is highly personal, and everyone needs to define their ethos in the context they are in. For a bank that is not stock listed, it may be easier to redefine its ethos because of the fewer parameters, but this shouldn't be an excuse for global institutions to hide behind.

Global banks have much less room to manoeuvre. "Much less" also means that is room left". Instead of hard exclusion strategies, but banks will often favour a transition strategy. They will not stop investments in, for example, fossil fuels but will steadily reduce it, thereby guiding their clients to a Green Transition.

One of those examples is HSBC. Their European Head of Sustainability and Strategy, Niccolo Polli joined us on stage a few times to talk about a new era for banking. He explained how the sentiment towards banking changed over time. He took the stage and was really honest about what went wrong.

To illustrate his point, he often refers to how Hollywood has depicted banking over the years, as he believes that films offer a good reflection of the public's perception of banks and their operations. Films like Trading Places and Wall Street in the '80s showed how wealthy bankers were, somehow detached from the average person's reality, though not inherently villainous. The 1990s portrayed bankers as less ethical, with movies like Rogue Trader and Barbarians at the Gates. The bad guys were still people, not institutions. After 9/11, Hollywood shifted its focus temporarily, but the Global Financial Crisis brought everything into focus. The financing of terrorism leading to 9/11, the extensive job losses, and personal turmoil post the global crisis, as well as the unethical conduct at some financial firms, inspired a new wave in Hollywood, he explained. "Too Big To Fail", "Margin Call", "Arbitrage", and "The Wolf of Wall Street" highlighted a lack of ethics. On top of that, series like Dirty Money showed life at a bank behind the scenes, and it wasn't anything to be proud of.

Niccolo shared: "Trust plummeted to around 30% and has yet to improve to pre-crisis levels. Only 70% of people say they trust their banks to do what is right, and only 35% of adults are confident that banks can keep their money safe (US statistics, March 2023). That has to be the most basic thing that you ask from a bank: to keep your money safe! So, today, I see a tremendous opportunity for the industry to change its ethos, to change its culture and public perception. We are currently in a global crisis, and that is climate change. Scientifically speaking, you must go back 3–4 million years when the world was hotter than it is today. We know that there is a direct link between carbon dioxide in the atmosphere and planetary temperatures. We also know how to slow and/or stop global warming, and we have most of the tools and technology at our disposal to do so. And banks have a role to play in making this happen".

He admitted that banks have to take responsibility but quickly added that waiting for regulation to come into place would be an incredible waste of time. HSBC is committed to achieving net zero in its operations and supply by 2030 and its financed portfolio by 2050. Net zero is the ambition, meaning that HSBC, or its clients don't need to have zero exhaust, but whatever is left needs to be compensated by carbon credits. The 2030 target is achievable for at least 80%, he said, but this is much less certain for 2050 (likely closer to 50%). So, the way they approach this transition is to first look at their own footprint. Then, they will identify sectors that matter in the Green Transition and establish a pathway for each specific region.

Then, they assess individual companies in those sectors by comparing their transition plans with HSBC's own projections. Companies that perform better than expected and help HSBC transition faster will be incentivised with better pricing or product conditions; others will see a reduction in their financing. With the current technology stack, an estimated 50% of 2050 can be achieved, so an important 3rd pillar to become net zero by 2050 is driving innovation. Banks play a fundamental role in reallocating financing to sustainable innovation. They can guide funds in the right direction. They may not be the biggest risk-takers, but they have the right tools and network to bring private equity in contact with bright minds looking for money.

In January 2024, they announced that their sustainable transition strategy is based on three axes:

- Supporting an enabling environment, working closely with the public sector, industry, and civil society to support an effective and standardised climate policy
- Mobilising market participants, like clients, investee companies, and delivery partners to accelerate transition solutions and journeys
- Transforming themselves, engaging both employees, suppliers, and investors on the road to net zero (Thompsett, 2024).

"Well-functioning and well-capitalised banks are fundamental to help economies and societies grow by providing funding to people and companies that can use it most effectively. And now we need to transition world economies from carbon to electrons, and banks will be fundamental in making this happen", explained Niccolo. That is also why his talks were titled "a new era in banking": "We need to live up to the expectations placed on us and also take advantage of this opportunity not just for the bank, but also for our civic responsibilities. This also allows us to rekindle our purpose as a bank to society at large" (Coeckelbergs, A new era in banking: How did we get here? An interview with Niccolo Polli—Head of strategy and sustainability at HSBC Europe, 2023).

Corporate loans are the big tickets for banks to activate deposits on their client's accounts. Although every initiative towards a more sustainable future is valuable, it is worth noting that the impact between corporates is very unevenly distributed. Since 1988, for example, 100 companies generated 71% of the global industrial greenhouse gas emissions. Of the global industrial greenhouse gas emissions between 1988 and 2015, 51% came from only 25 fossil fuel producers (Carbon Dislosure Project, 2017). Big corporations need a lot of funding to grow and to renew their business. Banks can determine which clients to serve, how to serve them and under which circumstances.

The challenge for banks brings us back to the ethos and the rhetoric triangle at large: people don't always believe banks when they share their good intentions in changing the world. It is too slow, or the impact is too little, or banks only do it for the profits. People have lost faith in banking, and they insufficiently understand the impact of global warming on the one hand and the transformative forces that are triggered by the Green Transition.

You see, in this debate, people who have lost trust in banking do not believe in their good intentions (ethos). Due to the complexity of the problem, combined with the complexity of banking itself, the aspect of logos is compromised. Moreover, it is imperative not to overlook that the Green Transition is one that impacts and even harms everyone, thus turning it into a highly contentious debate where individuals readily adopt a defensive stance. When individuals are defensive, they become more challenging to persuade, and the emotional connection (pathos) may become severed. Therefore, banks must carefully consider how to communicate their sustainability intentions in a manner that is clear, thorough, and coherent.

Niccolo Polli didn't hide any of the bad behaviour, by the way, and acknowledged how wrong some of the actions from the past were. That demonstrates leadership and humility. Ethos…

Throughout the years, there have been many cases against them. And yet perhaps that is why they now bring such a compelling narrative in sustainable banking. Completely different from vdk bank, they are a global institution, which makes it much harder to make harsh decisions on investment strategy, but they committed to having their operations and supply chain net zero by 2030, and they have a strict plan to have their entire portfolio net zero by 2050. That plan is built on scientific insights and mathematical logic. Logos

The plan is made in dialogue with the client, who needs to prove they commit and keep committing to that plan. If they perform better than planned, they get access to more investments; if they perform worse, HSBC will cut their investments. Pathos

In this respect, ING Bank recently announced that it scored its 2000 largest customers and asked them for a plan to reduce its impact on the planet. If these customers have not done enough by 2026, ING will end the relationship with them (Serrure, 2024).

8.5 Consumer Loans and Mortgages

The obvious domains for banks to contribute to a greener, less fossil-fuel-dependent earth is through their investment and corporate loan strategy, without a doubt. The clearer and more transparent that strategy is, and the more it is translated into the dialogue with customers, the more effective banks can execute those strategies. This plays in both directions: the deposit side, explaining how the money will be used, and the investment and lending by setting clear conditions to sign the deal.

We have another significant category that is responsible for 60% of global emissions: household consumption. When you think about this, do you also consider the fintech organisations that have developed carbon calculators to measure the carbon footprint of consumers' purchases? Or companies like Alipay that have implemented tree-planting initiatives as part of their loyalty programmes, turning consumption into a game for a better world? While these efforts may have contributed to creating new forests and influenced people to change their behaviour to some extent, I still see them more as a marketing play rather than a significant catalyst for change. These initiatives may enhance a bank's public perception as environmentally friendly, but they won't fundamentally reshape its values into concrete and significant results.

The true impact will only be felt if we focus on reducing the carbon footprint of housing itself. This is where retail banks can prove their value for a sustainable future, but it is a difficult path for several reasons. Many people may not be motivated to insulate their homes because it is a long-term investment that addresses a problem for future generations which may not directly affect them now. Moreover, even with government subsidies, the time it takes to see a return on these investments is often too long for individuals who have to prioritise their immediate needs, like food or school expenses for their kids, over making a costly contribution to a better world.

Another issue is the large rental market. Many people rely on landlords who view their real estate as a pure investment rather than a long-term dream. If new investments, such as insulation, do not immediately add value, there is minimal interest in putting money into them, regardless of whether the people living there like them or not.

Retail banks that want to prioritise sustainability and redefine their values face a complex task. Simply offering competitive loan rates for eco-friendly home renovations won't be enough. Offering a better rate can incentivise people to invest in lower carbon footprint, but remember: the provision of a better rate for sustainable investments is just a way for banks to protect themselves from their mortgages becoming stranded assets. The better rate from the bank is as much an investment in the future for them as the renovations are for the client.

Banks will need to facilitate sustainable education and ensure that their customers understand the impact they can have. They must engage them on the path to net zero and ensure that every consumer leaves a mark on that path.

In the unequal world we live in, banks that truly care about the planet need to be more creative. Banks have a choice to serve everyone in the best way their situation allows them to, risking discrimination and inequality. In a way, this inequality cannot be avoided, as it is inherent in the risk management role banks have. However, they also have the means to keep that inequality to a minimum, creating a sense of solidarity between those who have capital and those who need it.

How can banks facilitate a transition without harming the living conditions of clients who already experience daily financial stress? Currently, regulations limit the options for banks in this regard, as Guy Schellinck, CEO of Beobank, explained: "Those who are less fortunate and have limited resources are more concerned with meeting their basic needs on a daily basis. They cannot afford to think about the future and invest in it, unlike those with access to capital who can plan for the long term. We need to involve everyone in the transition to sustainable energy and investments. This can be done by making them pay the same amount for energy and investments as they would if they did nothing. This would create cost neutrality and help build a future-proof product in the medium term. However, implementing this is difficult due to legal restrictions, among others".

Stimulated by supervisors, banks traditionally have collateral and assess their long-term value based on historical results, explained Roland van der Vorst, Head of Innovation Wholesale Banking and Rural at Rabobank, a few years back. "What you need to do is actually do it the other way around. You need to translate future value into today's benefit", he explained.

That is also the kind of solution Beobank is looking for in Belgium for the real estate market for exactly the same reason.[1] Less fortunate people may well understand the value of investing in more insulation, but if the investments are so big, it is simply impossible for them to make them. Beobank hopes to find a solution by solving this challenge with one of the major energy suppliers (Coeckelbergs, The future role of banks in our society—Interview with Guy Schellinck, CEO, Beobank, 2024).

In the Netherlands, I found good example of how to deal with the Green Transition in an unequal world while avoiding heavy lending regulations. As a cooperative, Rabobank has the advantage that not all profits must go to shareholders. In 2023, for example, they allocated 400 million to Cooperative Sustainability Initiatives. One of the more surprising initiatives was to allocate 1.000€ to all mortgage customers with houses with energy labels F or G to insulate their houses. In doing so, they stimulated people to invest in insulation, knowing that less wealthy customers pay less for the total investment and have a higher portion paid by the bank. The bank allocated 128 million for it (Rabobank, 2024). It costs them money, but the return is likely higher on the long term, and it contributes to the cooperative commitment they made to their members and clients.

Another great initiative is Rabo Smartbuilds, an ambitious initiative to build 12,000 modular and sustainable houses for the rental market in 10 years. These high-quality and energy-efficient houses are often built in locations that don't allow permanent housing. By making them modular, these houses can be moved to other locations if needed. In doing so, Rabobank contributes to both the challenge of a shortage of houses on the rental market and the path to net zero. On top of that, they work with the local communities to ensure that the housing also contributes to a stronger local anchoring by focusing on specific segments that get priority access to these houses (Rabobank, 2024).

Rabobank is doing much more to reach net zero, but they are not too proud to share more can be done. In their report "Our Impact in 2023 – Better World", they let one of their clients, Stijn Vos, CEO of Enstall, share: "In America, we're now also approaching target groups who are resisting the move to sustainability. For instance, people who have less money to invest in solar panels or who rent their homes. The question we're asking – and which you should be asking as a bank – is: how can we activate these people? Bearing in mind that this is a large group of people, and we really need them to make the energy transition successful. Someone has to come up with financial solutions, and there's probably a lot more Rabobank could do on that" (Rabobank, 2024).

As much as this applies to Rabobank, it applies to other banks as well, and that is exactly what a bank's mission is: to redefine its ethos by focusing on sustainable transition.

Beyond the provision of financial services and financial support, banks have another valuable asset to build the road to net zero in housing, and that is their network and data from mortgages. Several banks are offering tools and platforms to help their customers. Those customers can be consumers who need support to understand the best next steps and best pricing for their renovation projects on the one hand, and the bank can match them with professional clients that provide renovation services on the other hand. Organisations like Setle and home Qgo can be helpful partners in between to help with real-time insights to consumers on how to make their houses more sustainable and at which estimated cost.

Through platforms like jaimy, founded by Belfius Bank, I found a painter to redecorate my house in the past, but the platform can equally provide me with any other renovation service. Banks have a strong network, and by bringing all these parties together, they not only add more value to their service offering, but also strengthen their customers, whether that is through increased revenues or more solid real estate.

8.6 Conclusion

The Green Transition is a tremendous opportunity for the industry at large to demonstrate their role in society today, and in transforming for a better future tomorrow. Regulation alone will not be enough to prepare earth for future generations, it also requires financing, and a lot of it.

Whether banks are purely looking for attractive rates and returns or want to build a meaningful business for their clients, society and the planet, they will be boarded on the train of Sustainable Finance of an estimated 150 trillion USD globally by 2050.

Banks must ask themselves how they wish to join that train. How big do they want their impact to be, and how can they achieve this? Universal banks may have larger leverage than small ones, but the consequences of their actions and responsibilities are also much bigger.

We don't know today what the path to tomorrow will look like; we only know it will be demanding to live with a small carbon footprint on the one hand and to restore what was broken on the other. That means banks will need to invest in change and innovation. Today's revolution is not just a new development. No, we'll need to transform all that has been built before. That is a tremendous change compared to the Industrial Revolution, which, simply put, added building blocks instead of replacing them.

Today, we see, feel, and experience the impact of global warming, and it isn't warm and cosy. People understand things need to change, and banks should grasp the opportunity to assist in this change one way or another. Some can afford to do this more progressive than others, and that is OK.

Every bank must reflect on how they internalise legally required new procedures, reporting, and customer insights. This is an opportunity for all to rewrite their ethos and demonstrate that they are serious about helping society through this change. Additional questions to understand clients' investment appetite is an opportunity to better understand the customer, to build more compelling products and services and to strengthen the messages the bank wants their clients to remember.

vdk bank was a good example of how a bank reframed its message in a very tangible way. They made it easy to understand what an impact a bank can have with people's money. The only thing they needed was a coin, with three sides, to symbolise people, the community, and the planet.

Increased risk weights for carbon-consuming assets and related stress tests are an opportunity to review the bank's balance sheet, but also the pricing model, and negotiations with clients to help them transition to alternatives that consume much less carbon. That is where logos and ethos come together, but it isn't a walk in the park.

Most banks still struggle how to rebalance their pricing model, anticipating changes in valuations of their assets. How much negotiation power do they have today, bearing in mind the valuation of investments in the future?

That thin line is what HSBC is walking for their big ticket corporate loans, and many other banks with them. But also in retail banking, this is a daily board discussion because the risks are new, and the valuations were never seen before either. Banks were used to look at the past to make predictions for the future and that methodology is no longer enough.

In that respect, simply distributing money to invest in insulation, as Rabobank did, made sense. It stimulated people to invest in their housing, future-proofing their homes and, as a consequence, also their mortgages, which affect the bank's balance sheet.

"Leaders who understand that ESG/sustainability is much bigger than a spreadsheet or a market segment. Those leader banks with purpose at the forefront will be as big a change to finance as Amazon was to shopping. If you can't see that, you are going nowhere in banking – the future is ethical and digital. The planetary balance sheet is in the red and has to be reconciled. And we're going to need banks to do that – in fact I think, and it's been said a few times, that the finance sector will save the world in the end. That's the ethos we need – but you've got to build it with your heart and as you say – pathos", Corin Millais (Coeckelbergs, About David, Goliath and The new ethos in banking, 2023).

Note

1. Rabobank solved the problem of reaping instant benefit from investments in restoring natural capital by working with carbon credits. Acorn (https://acorn.rabobank.com/en/) is one of the projects using this technology. They set up a business in agroforestry, where small farmers are rewarded for planting trees. These trees improve the quality of soil, capture water, improve biodiversity, and what's more: they capture carbon. By monitoring the delta in biomass, they calculate the amount of carbon that is being captured and that is being sold to big companies who wish to be climate neutral (Coeckelbergs, Rabobank's Global Food Revolution, 2022).

9

The Road to a New Ethos in Banking

9.1 Introduction

Instead of greed, complexity and indifference, banks should be standing for prudence, accountability and accuracy. As one executive shared on stage at one of our conferences: "We need supervision and regulation. But more than anything else, we need values. That means that people who work in the banking sector need to live up to the values of service, humility, society, and risk management. Do they need to be entrepreneurs? They need to understand entrepreneurs, but they shouldn't be them as we saw with SVB bank. The moment I wear new sneakers, you should immediately mistrust me. Because a real banker is, I would argue, a little bit boring".

I'm not trying to define "boring" here, but I guess you see the point this person was making. Banks are risk managers, whether that be operational risk, credit risk, climate risk, or any other risk. Of course, you have ambitious and innovative financial institutions, and rightfully so, but there is a reason that often the innovation process goes slower than elsewhere. Banks play with other people's money and that comes with a responsibility. If they don't take that seriously, they risk handing over the losses from ignoring those responsibilities to society, which is the complete opposite of what banks should demonstrate in defining their new ethos.

I wouldn't call any of that boring. If banking is about realising dreams, I believe banking is a very inspirational industry. Banks facilitate daily transactions, all over the world. Banks help you to buy a house with the money you earn in the coming 10–30 years, or even more. Banks are the support line for entrepreneurs and businesses in case of liquidity issues, and they enable

R. Coeckelbergs, *A New Ethos in Banking*,
https://doi.org/10.1007/978-3-031-78234-3_9

the necessary investments to grow. To ensure more people understand this, many banks will need to rewrite the narrative in banking. I hope this book has shown that this can be a challenging but very realistic ambition.

To ensure you understand how to travel the road to a new ethos in banking, I set a couple of milestones in the right direction. I also have questions because I cannot define your new ethos; each organisation has its own exercise to do. After all, ethos is highly personal.

9.2 What's Your Calling?

"Every bank must ask itself: what is my calling as a social institution, and where can I have an impact? Each bank, whether small or large, needs to understand what it is that they can influence and have an impact on. What is the issue or topic they can work on?", Martin Rohner shared with me. Martin is the Executive Director of the Global Alliance for Banking on Values, "a global movement of frontrunners in sustainable and value-based finance". Created by the CEO at the time of Triodos Bank in the Netherlands, Peter Blom, still today they are an example of the impact value-based banks can have across the world. They have 70 members today (Coeckelbergs, The new ethos in banking: Values-based banking—Interview with Martin Rohner, 2024).

You don't need to be a member to understand the importance of the question that Martin raised here: what is your calling? As an organisation, it is fundamental to understand and live the values that you preach because these should be more than words. Whether it is to offer the best possible digital experience, the most invisible and frictionless services, or the most sustainable banks, there must be something to have an impact.

That impact should not just be an organisational differentiator but also the glue between the organisation and its people (whether employees or customers) and between the employees. Your calling should be a group mission to do better in the future. That is what a meaningful transformation is about. That will help understand everyone in the organisation for which bigger reason they wake up in the morning and leave (or simply connect) to work. Embracing ethics and values for a meaningful transformation requires a common understanding of meaningful work. Eynikel found that work becomes meaningful when:

- You experience that what you do corresponds with who you are;
- You experience that your work has a visible impact and contributes towards the larger whole of your team, organisations or wider society;

– You can see that your organisation is striving to achieve meaningful results that are aligned with the group values;
– You feel valued and appreciated within the larger group in which you work.

This inherently means that meaning is about connection. It is about connecting team members with the team, the teams with the organisation, and the organisation with the world.

If you believe this is a pure HR topic, think again, because this alignment has a direct relationship with the results of the company. Philosopher and management author Danah Zohar spoke in that context about spiritual capital, a third form of capital next to economic and human capital.

When meaning is well understood within the organisation, and everyone is aligned in the direction the organisation should go, spiritual capital will thrive. Besides good team spirit, higher levels of spiritual capital will motivate people to do their best, not just in good times but also in bad (Eynikel, Check in—The search for meaning and purpose in business, 2021). It is the extra ingredient to embrace values for a meaningful organisational, digital or business transformation because it matters. It matters for the customer; it matters for your employees and it matters for the organisation.

Banks that embrace ethics and values for a meaningful transformation will have stronger team cohesion, willing to go the extra mile beyond their personal interest. McKinsey refers in this respect to the term "Meaning Quotient" (MQ). When MQ is low in a business environment, employees put less energy into their work and see it as "just a job" that gives them little more than a paycheck. They found that people in a high-IQ, high-EQ, and high-MQ environment are five times more productive at their peak than they are on average, showing the immediate impact on company profits in building a sense of genuine purpose.

According to them, inspiration stories should circle around five layers: the company—society—the customer—the working team—themselves. McKinsey found that in a cost-reduction programme of a large US financial services company, the moment they restructured their change programme and built the motivation story around those five layers, instead of simply explaining the rational change story, the share of employees reporting that they were motivated to drive the change programme jumped from 35% to the majority 57% of the employees (Cranston & Keller, 2013).

Consciously translating the bank's mission statements into its actions will make the organisation and its message to the outside world more reliable and, in the long term, have a higher trust factor. But that is not enough to redefine a bank's ethos because a bank doesn't work in a sterile environment.

9.3 Who Do You Serve?

During the desk research for this book and after some of the talks and interviews I had with executives, I sometimes wondered whether the writing of this book would have any impact. Who cares about a new ethos in banking? Why should decision-makers listen to me? After all, not only do they know better, but most of them have up to double the experience I have, no?

I remember a conversation with a bank executive who complained that journalists only care about the numbers. They can organise press moments around side activities that showcase their contributions to creating a better, more inclusive, and healthier society as much as they want, but such events will rarely attract as much press attention as the release of new financial results.

This brought me to a second doubt but also a kind of reassurance. The press wants to hear about the numbers, assuming that is what the audience wants to hear. The complaint that the press is only interested in that reassured me that talking about ethos is more relevant than ever. Because there is much more to be said, unfortunately, it lacks an audience to listen. A strong ethos will make people curious about your contributions to a better world, and maybe more importantly, it will make people understand that banks need, but also serve, more than only loyal shareholders to thrive and survive.

A strong ethos inherently means that the bank has a more captivated audience, willing to listen.

"Banks should not primarily focus on shareholder value. Depositors have much more money on a bank's balance sheet", Nout Wellink, a former Dutch central banker, once said. Without customers, there is no liquidity, and without liquidity, there is no financial intermediation and no bank whatsoever.

Wellink was criticised for that point of view before the GFC because it was not aligned with the common capitalist sentiment back then, where shareholder value was the mantra for most organisations. Ultimately, though, a solvency crisis and a liquidity crisis will often have a cause-and-result relationship, not always in the same order. Wellink's comment may have sounded outdated back then, but it was not too far off in his vision of banking. Ultimately, a bank has no chance of survival if it cannot keep its customers happy. Banks need to balance the interests of their stakeholders, which include shareholders and customers.

Those interests can be very different. That diversity is a blessing for the industry. In some cases, customers are the shareholders, in some cases, banks work for members and in most cases, there is a split between shareholders and customers. Many waves of consolidation have reduced this diversity in business models.

Exactly that triggered the authors of "Virtuous Banking – Placing Ethos and Purpose at the Heart of Finance" to say that one of the action points to make banking virtuous again is promoting industry diversity. We need more diverse banks, each with their unique business models and service offerings. Although this isn't a measure a bank can implement on its own, it's worth noting that supervisors also play a role. Throughout the world, the banking industry has always had various types of banks. Over the last decades, this diversity has decreased in favour of economies of scale. Somewhere, the world has lost the idea that the Board of Directors, the ones that oversee the executives, must defend the interests of the bank, not just the interests of shareholders. That means that they must respect the unique model each bank represents, also if that risks playing against the classical slide decks at conferences.

Increased competition will promote better customer care and protection. Regulators have several measures in place to safeguard customers, and the industry has made many improvements concerning its duty of care, but more improvements are possible.

The need to scale in a transactional industry like banking is clear. However, we also know that technology costs are going down, meaning that the rule of economies of scale is slowly but steadily losing value. New competitors who embrace the latest technology to set up new ventures will be the first to benefit from that, which will trigger a more inclusive society and is a clear call to action for the existing industry players.

Tom Dehaene, Director at the National Bank of Belgium is very clear by the way: the narrative that profitability of a bank depends on its size is not entirely correct: price setting is a much more important element in a bank's profitability (Suy, 2024).

But also diversity in banks remains a common concern. If banks want to be good servants of an entire society, they need to be a mirror of society that understands and feels how that society experiences banking and the need for banking. Clearly, the most spoken aspect is one of them: banks need a diverse workforce to capture all the needs of their target audience. For too long, bank leadership has been, and too often still is, dominated by "grey, white men". Society flourished through freedom of speech and expression, but banking didn't follow.

Banks must understand that they are built on a perception of understanding their customers, just as customers build their trust on a perception of understanding their bank. To come back to the book's thesis, if the banking industry wishes to restore pathos, it must encourage and actively seek more diversity across the organisation, in line with the target audience it wishes to serve.

That is exactly why rewriting the narrative and finding that new ethos in banking is fundamental for the future of the economy and society at large. Most banks are not just profit-making machines. They have a bigger sense of existence, and every employee must understand the values behind that and live them throughout the day in the services they offer to customers and in the products that they build.

As a bank you must understand that you serve your shareholders, and also your employees, but first and foremost your customers. If you wish to move from transformation to a meaningful transformation, it is fundamental to define meaning to your organisation and the people that are being served. That is what the new ethos in banking is about: it is the result of a good understanding of expectations (pathos) and building compelling products and services to live to those expectations (logos).

You need to know your customer's worries, aspirations, expectations, and dreams. You need to know how your customer wishes to bank and deal with his financials and understand his/her struggles in life. That is the only way a bank can effectively help in realising dreams.

You also serve the community and the planet your customers live in. How you balance this with your shareholders, employees, and other stakeholders is up to every organisation to decide. Just make sure to never forget that bigger picture when redesigning or defining your new ethos.

9.4 Shareholders Are Not the Enemy

Clearly shareholders are important stakeholders, but they are not the only ones. You need to keep all the stakeholders happy, and keeping someone happy starts with setting clear expectations. Being stock-listed, doesn't mean it is impossible to move beyond the traditional money-making machine that banks can be.

Throughout this book, we discussed industry initiatives to promote financial and digital education. Countries that see a sharp decline in cash take action to promote financial inclusion and access to cash. None of that will directly contribute to share price or dividends, but these are fundamental steps to make banking a more trusted industry again, one that considers the doubt and worries of its stakeholders, and that ultimately leads to be a better appreciation of the industry overall.

Many banks invest in micro-lending initiatives and innovation accelerator hubs. These are longer-term investments that shape a bank's identity. Shareholders must respect this and acknowledge it as part of the company they invest in. So, no: shareholders are not the enemy of a new ethos; they must embrace it with care. A trusted brand is a resilient brand, and that will lift the share price as well.

That doesn't mean it can be challenging to balance shareholder value and ethos. In that respect, it is worth looking at Erste Group. That bank remains loyal to its original mission, although by now it has grown to be one of the largest financial groups in Eastern Europe, with a balance sheet of +€300 billion and stock-listed. This Austrian bank group operates in 10 markets in Eastern and Southern Europe, including corporate, retail, private, and wealth banking.

The group was founded more than 200 years ago by a Catholic priest and a few nobles with the mission to democratise banking in an area where banking was only available to the wealthy. The democratisation of banking remains in the DNA of the bank. Therefore, Erste Group Bank and ERSTE Foundation set up a joint venture called Erste Social Finance Holding, to grow and provide access to financial services to all. You read this correctly: it is a stock-listed company, and they have a separate business unit to focus on the democratisation of finance.

Peter Surek, Head of Social and Banking and CEO of Erste Social Finance Holding, explained that there are three fundamental principles related to their activities:

- they must have a societal impact;
- they must be linked to banking but separate from the existing major business lines;
- they must be financially sustainable.

"We should cover our costs. The Board understood that our ROI target could be lower than other business lines, but we had to cover our costs at least; we are not a charity. That was the task which we got at that time", Peter explained. It led to 3 segments where social banking could fill a gap where regular banking was failing:

- Micro enterprises that required a microfinance proposition;
- Non-profits, NGOs, and social enterprises no bank was willing to finance;
- People at risk of poverty or financial exclusion that require a basic bank service and access to credit for a better (financial) life.

Although these initiatives are only a small portion to the group's total assets, according to Peter, he believes it has a lasting impact from both a financial and a reputational point of view: "You can compare us sometimes as a kindergarten, or a health centre for the bank, for the clients who are not yet ready for regular banking. We are catering for them; we are developing them. Once they are doing fine, they will be a very loyal client of the bank. Because if you help somebody in a difficult situation, this customer is usually much more loyal than those we get by some marketing campaign" (Coeckelbergs, Leveraging finance for social impact—Perspectives by Erste Group, 2023).

Building a business on purpose does not contradict making a profit or creating value for shareholders. It is all a matter of expectations. Martin Rohner, Managing Director of GABV, explained in that respect: "The focus of value-based banking is on achieving a positive impact on people and the planet. What I observed is that even though our members are not primarily focused on profit, in most cases, they are actually very successful and profitable. They might benefit from higher credibility and higher customer loyalty, which makes them more resilient to crises, etc. So values-based banking pays off in the long term and makes banks more stable" (Coeckelbergs, The new ethos in banking: Values-based banking—Interview with Martin Rohner, 2024).

9.5 A Flavour of Magic

How can you make a difference within the context of your work? How do you respond to human, moral and societal needs in a way that matters and that is meaningful to the ones around you, whether that be around sustainability, promoting entrepreneurship, inclusion or any other challenges we face as a society today? In which direction does your moral compass point?

Defining ethos starts with retrospection because, after all, it is not something you own. "Ethos is not so much about ethics as it is about character, and in the context of banking, I would say culture. So, if you say we need a new ethos in banking, to me, this refers to a cultural change that is required in the industry", explained Niccolo Polli, Head of Strategy Europe at HSBC (Coeckelbergs, A new era in banking: How did we get here? An Interview with Niccolo Polli—Head of strategy and sustainability at HSBC Europe, 2023).

Your retrospection will make you understand how your organisation is being understood and perceived. Changing that ethos is difficult but not impossible, as the many examples in this book show. Redefining ethos requires organisations to clarify how they wish to contribute to a better planet, which values they want to stand for and which purpose they wish to fulfil.

More important than that is to live those answers across the organisation. It should not be a top-down manifest but one that is being respected by board members as well as support functions and everything below and between. That is the only way to break with the kind of "magnanimous speak" you preach by announcing your new ethos.

Michael Anseeuw, CEO of BNP Paribas Fortis, provided me with the last piece of advice for the industry to improve its narrative in banking. He explained that bankers should be more like Mary Poppins. Mary Poppins effaced herself for the children. "It was never about her; if was always about the children", he said. "Banking is not about ourselves, but more about our customers and society at large".

Mary Poppins used magic to surprise these children and add a spark of brightness in their lives. Remember, banking is about realising dreams. Banks need to use more of that magic to make dreams happen. They allow you to pay by card or phone anywhere in the world. They allow us to buy a house today with the salary of the many years to come. It is about magic. In this context, the ethos of a bank is about understanding the bank's core magic skills and applying them to make the world a better place.

Mr. Anseeuw added one more reason to get inspired by Mary Poppins. Banking became too serious: "A last element is that Mary Poppins always comes with a tongue in cheek and a touch of humour. We bankers became too serious. We can blame ourselves for that, but we are also pushed into that role. We are expected to be very serious. So, let's bring back that touch of humour. Let bankers be a bit more Mary Poppins again" (Coeckelbergs, About Mary Poppins and the new ethos in banking, 2023).

9.6 The Power of Consistency and Clarity

For this book, I was recommended to read a book "Perfect Prey" (originally published in Dutch as "De Prooi") a novel that describes the rise and fall of ABN AMRO Bank. Certain moment, the former Chairman of the Board of ING Group, Cor Herkströter, is being quoted: "It is fundamental that the principles of an organisation never change. Principles that change every year become worthless for employees. In addition to clarifying the purpose of an organisation, it is one of the most important hobs of a leader to define and keep control over the values of an organisation". Consistency will make a behaviour, a message, or an action reliable. It shows that a person or an organisation is serious about it.

Strategies can change and must change. Because a strategy defines how to respond to a changing environment, the DNA of an organisation, the behaviour and its values must be rock solid in organisations that pride themselves on having a strong ethos. The power of consistency means that the ethos can survive a CEO switch because Supervisory Boards will ensure that the organisation remains loyal to its ethos and moral duties.

That also means that it is incredibly difficult to drastically redefine the ethos of an organisation. Trust leaves by horse and comes back on foot, and the same is true with ethos. Rewriting a narrative in a way that changes an organisation's DNA so that people feel, breathe, and live that DNA as well, from both a leadership and employee perspective, is a multiyear track, and sometimes even longer.

To make it more complex, this is only the start because, in the end, ethos belongs to the audience, as we explained. That means that the change in the leadership and the way that employees work and serve customers must also be felt by employees, the people involved, and the stakeholders.

Be explicit about the values you stand for and embed them in a transparent governance model that is being accepted by the entire organisation, from top to bottom. Set the right expectations to everyone involved and live up to those expectations.

As a bank, you can say you take financial well-being seriously. Still, that message loses all its power if the same bank starts aggressively issuing high-margin loans to high-risk people because the results are under pressure. Implementing a fancy personal finance dashboard in the mobile app will not change that because the bank's actions were not aligned with its behaviour.

As a bank, you can say that you care about the planet and a greener future, and as much as that message adds to the importance of the theme, it is not enough if that same bank keeps unconditionally investing in fossil fuel companies. The argument of "if we don't do it, others will" will not change that. If that bank wants to rewrite its narrative and build a new ethos that prioritises contributing to a greener planet, it must act in line with that mission, in one gradation or another. I will not claim that this must be done through an exclusion strategy or transition strategy, which is for the bank to decide, but a consistent strategy must contribute to a greener planet. Otherwise, the message gets waved away for being greenwashing, damaging the already poor reputation the bank may have.

By embracing values and ethics for a meaningful transformation, banks will be able to find a new ethos, or perhaps refine their ethos, and end up being more trusted partners to their clients and all stakeholders involved. As we know, trust comes by foot and leaves by horse, and if you are up for a long trip, cadence is the keyword. The rhythm of going, foot by foot, mile by mile, will bring you everywhere. That is what got me to the finish line of a marathon twice in the past. I would count down the kilometres and keep going consistently.

Ethos follows the same flow, with the difference that it doesn't have a finish line.

None of this means that ethos should be written in stone. It is a highly contextual concept, not only on an organisational level, but also in time and geography. Like a marathon requires constant evaluation of how to reach the finish line and how to consistently count down the kilometres, also ethos will evolve.

The value of creating safe environments made banks decide to no longer invest in the weapon industry, for example, but that same value is being used today by many financial institutions to invest in them again because of geopolitical tensions. These institutions remain consistent in their values, but the translation to their day-to-day activities evolves.

The roll-out of GenAI tools today is going rather slow, because banks consider privacy, transparency, and accountability. Those same values may result much faster developments once we better understand its impact and limitations. These institutions remain consistent in their values, but the translation to their day-to-day activities evolves.

9.7 Time to Redefine Your Measurement of Success

The Enlightenment emphasised rationality, critical thinking, and individual freedom. One of the consequences of this was an obsession with measurability, manageability, and control. That may still explain why values like good care, human service, and societal support are still too often measured in numbers and lack every nuance (Eynikel, p. 81).

Banks, even more than other industries, need to conduct their business responsibly and have clear guidelines to set corporate governance and also clear guidelines to report these in the right way to the supervisor. That is not what ethos is about. Ethos is being expressed in the good behaviour of people in the organisation, and the common understanding of that behaviour, both today and tomorrow. Ethos is the collective responsibility, both organisational and personal and the actions that prove that this responsibility is well understood.

During my exploration of the new approach to banking, I had the opportunity to speak with Joris Krijger, AI and Ethics Officer at De Volksbank in the Netherlands. Banks need to be held accountable for the decisions that they make, he explained: "At the end of last year (2023), a report by KPMG showed that banks and insurance companies have the least trust when it comes to the use of AI and algorithms. Customers trust banks and insurance companies even less than the government with these technologies. I believe it has a lot to do with the opacity but also with the lack of accountability mechanisms when things go wrong".

He shared an ambitious goal for banks, emphasising the need to reassess how success is measured and the concept of social responsibility. What is a bank's responsibility towards consumers, and which role does a bank play in responsibly deploying new technologies to these consumers? Can banks afford to limit, for example, the usage of AI purely for commercial reasons, or should it be deployed for long-term improved financial well-being of the customer? And if so: how, and more importantly: how will they measure the success of it?

It is a question that requires a broader scope than purely the technological side. The accounting balance sheet has been invented in the fifteenth century. That is how long the shareholder model exists. Mr. Millais spoke about the triple bottom line, where next to money, also people and the planet must be considered, similar to the three-sided coin of vdk bank. That tripe bottom line has known a much shorter history, only around 20 years, he explained. It is a premature concept that shows change is possible, but we are not there yet. He believes that mutuals and other membership-based business models can lead by example (which again demonstrates the importance of diversity in the industry) (Coeckelbergs, About David, Goliath and the new ethos in banking, 2023).

Each bank must ask itself how it can meet societal expectations, not just today but also tomorrow and in 10–20 years. That is a serious challenge, and since banks are private organisations that must also make a profit to survive, asking these questions will require more than simply looking at the fiscal bottom line. Joris believed the measurement for success must be much broader if banks wish to be taken accountable for their societal footprint.

This deviation from the standard evaluation mechanisms will be difficult, but we are starting to see a change. Banks are taking more responsibility and showing transparency. A good example of that is yearly reporting on the road to new zero and other ESG initiatives.

According to Joris Krijger, this is just the starting point of something much more ambitious. "Right now, banks are punished for serving high-risk clients because that will not reflect well on their financial performance. But perhaps you are doing a good job in terms of social performance", Joris explained. "As we don't have any measurements for that now, or it's not agreed as something that determines the value of an institution, I think that's where the rules of the game conflict with the common good. I would like to see more of these measures as part of what defines success, to make sure the rules of the game will shift or fit better with what is in the public interest as well" (Coeckelbergs, We conclude our talk with Ethicist Joris Krijger about the new ethos in banking (Part 2 of 2), 2024).

It made me think back about a presentation at one of our events back in 2019 describing Zweite Sparkasse. Founded in 2006, Zweite Sparkasse helps Austrian citizens to overcome their financial difficulties. It is one of the initiatives of the ERSTE Foundation and builds on the goodwill of 350 volunteers, retired employees of Erste bank. Their measurement of success? To enable as many customers as possible to re-enter stable economic conditions, or practically speaking: to enable them to become mainstream, and ideally, but not mandatory, Erste bank customers.

What makes this case extra relevant is that by working with retired volunteers, the group adds a sense of purpose and social commitment for former employees, growing their ethos beyond the organisation, and keeping costs low in fulfilling its mission. By doing so, they anchor themselves more strongly in the local community (pathos). These volunteers meet with clients and provide them with basic banking services, but more importantly, with invaluable financial advice that makes a world of difference for these clients who are under permanent financial stress (logos). Erste Group is building on the mission that they have been living for 200 years: democratising financial services, a great example of the power of consistency in practice.

9.8 What's Your Story?

Now it is up to you.

Transformation today must start with the hope for a better world, and we are just at the start of this evolution. An evolution that requires a new ethos in banking. An ethos that favours transparency and customer centricity, society, and the planet as a whole. Finding that ethos start with finding your story.

If you found your organisation's NDA, you know which values characterise your organisation, if you have your mission statement, and you know how to evaluate how meaningful you are in living that mission, it is time to build your story, your narrative. We learned that meaning is about connection. That connection should be between employees and the organisation and also the organisation and the outside world.

What will you do as a bank beyond profit generation for shareholders? You have determined what values are important to you. You also determined who these values are important for. If these exercises mean that you are lifting your path of transformation beyond peer comparison, you set the first steps towards embracing ethics and values for a meaningful transformation. Well done!

That is the first step to the new ethos in banking, which is about redefining your narrative. You will need to find your story, and that is much harder than setting values, mission statements, and a purpose. Finding your story goes beyond building beautiful slide decks.

As beautiful as the social posts regarding volunteering activities with colleagues are, I can only hope that, in most cases, there is an intrinsic motivation behind these initiatives. Similarly, we see more bankers in education programmes at school in an effort to drive financial literacy. All these things contribute to "your story".

It doesn't need to be so catchy or highly visible. Often, the best stories are the small, or even invisible stories that demonstrate the pride of people to work for their employer. It is a story that is essentially how individual employees and teams translate what is important in their behaviour. Klaas Ariaans of ABN AMRO Bank is a good example of that. By transforming his organisation into a network of self-organising teams, he allowed them to excel in customer service. People were given the chance to organise their work in a way that was way more efficient and effective, creating a better work environment for employees and customers.

People were capable of experiencing first-hand what the company values meant to their day-to-day jobs. People were collecting a shared story that resulted from a common corporate DNA. Klaas gave the example of complaints handling, a service with traditionally a very low customer feedback score. The management decided to let employees decide to compensate or solve the complaint, for any complaint up to 500€. This freedom resulted in an increase of payout for two months, but more importantly, it generated more transparency of what was going wrong. People started solving problems, and after 3 months, the results were overwhelmingly positive, with both employees and customers being happier about the complaints process

(Coeckelbergs, ABN AMRO Bank's sales transformation with self-organising teams, 2022).

Sharing these kinds of experiences will trigger more stories, and when these stories can be connected in a consistent and coherent bigger picture, you have a culture, an organisational DNA that is not just written on slides but one that is embraced by all people in the organisation.

We saw it in the example of the Dutch Banker's Oath, where some organisations asked their employees to take the oath in a more personal setting, in front of the client, or even in front of a mirror, making it more than just repeating words because the law expects it (Coeckelbergs, The Banker's Oath and the new ethos in banking (part 1 of 2), 2024).

Another good example is what happened at Danske Bank. They rewrote their narrative after the many scandals in the past. They went through the dust but took the opportunity to transform into a more value-based bank again. History has shown how a bank can stand as a national symbol, and they lost this 10–20 years ago. They redefined their ethos and embraced ethics and values for a meaningful transformation, one where transparency and sustainability are at the heart of their business.

In the context of the Bank's vision "Making More Possible", Danske Bank also did a very interesting explicit story experiment. In a "Squiggly Stories" event, members of the board and the executive committee shared squiggly career experiences to over 100 attendees. After the success of this first event, they did more of these events, because by sharing these stories, 95% of the listeners agreed that they had new perspectives on progressing in the organisation. It made people more flexible in agreeing on a future role, not just looking for horizontal career paths (Tupper & Ellis, 2024).

Sharing stories connects people with similar experiences; in this case, it can challenge assumptions that work against the ethos you try to establish as an organisation, making a stronger connection between people. Stories can help determine the behaviour and collaboration of the organisation as a community.

Bridging the personal stories with corporate objectives in a meaningful way will also define the company's identity and the directions for a meaningful transformation, whether you wish to excel in digital experience, become an impactful sustainable bank or wish to be the best in supporting local entrepreneurship and the local economy.

So my call to you: what is your story? How will you and your organisation contribute to a better life for people, society and the planet? How will you embrace ethics and values for a meaningful transformation?

9.9 Conclusion

Throughout this book, you have read what made the banking industry lose confidence and trust. After close to 20 years, they are still suffering from trust, having left by horse after the global financial crisis. That means that banks keep having a hard time explaining that they are more than money-making machines, explaining the fundamental role they have for people and the planet.

In this close to final chapter, I tried to provide a couple of high-level tips for bank executives to restore, define or redefine their ethos by embracing ethics and values to facilitate this transformation.

The very first thing to do is to make sure you know what you stand for as an organisation, and everyone in that organisation. They determine the DNA of the bank, but also the culture among colleagues, and the character of the organisation perceived by the outside world. Without it, is it pointless to figure out how to frame a transformation based on ethics and values.

In Short: What Is Your Calling?

Once you understand that, you will need to look critically at who you are primarily serving. The point made here was how important the client is for banks. This is all about Pathos and building a clear understanding of who you serve and who you don't serve, and perhaps also building a priority in case of conflicting interests.

That is where many may point to the role of the shareholder, but in my opinion there should not be a conflict between shareholder and customers, society and the planet, as long as the expectations to all are clear and reported transparently.

Every organisation has its own secret sauce, and a diversified industry is a thriving industry since different banks will inspire each other by demonstrating their unique tastes, their flavours of magic. What is your secret sauce, and how will you make sure you have the right recipe to let your target audience not just taste but eat your mission and the way you build a better world together?

Up to you whether you focus on digital excellence, sustainability or any other relevant topic. Ethos is highly personal, and this is your chance to show that personality to the world. Just make sure you are consistent and transparent in the ethics and values you embrace. You want to be remembered for those, as they form your identity, your DNA. The context may change in which you live the values, but if you have a solid and consistent story and are transparent in your actions, people will remember you for it. That is their contribution to your ethos, which belongs to the audience, not to you.

The proof of the pudding is in the eating, and for decades, we have worked with rational KPIs and numbers to rationalise and simplify measurement of success. In today's world, where capitalism is more than shareholder value, in a world where the usage of technology can add incredible value but potentially at the stake of tremendous damage for others, we may need to redefine our measurements of success. We may need to re-evaluate Logos and align it with today's new challenges and expectations.

Now it is up to you: what is your story? How will you translate all of this in your bank, making sure you contribute a fair profit for shareholders, and equally contribute to people, society and the planet. What is important to you? Why do you want to wake up in the morning and what makes you feel good going asleep in the evening?

How will you define the ethos in banking, embracing ethics and values for a meaningful transformation?

10

Fast Forward: "My Life, My Bank in 2040"

10.1 Introduction

2020 was a special year, for a couple of reasons. I'll remember it as the year that my first son was born. Those days I was talking to Chris Skinner, global influencer, commentator and fintech troublemaker. He told me about the new book he was working on and asked me if I wanted to contribute as well. It is about what banks can and should do to promote inclusion, diversity, and sustainability and what technology can do to create a better financial world. Chris asked me to write my vision on banking when my son would be 20, so banking in 2040. Looking back, it is probably one of the first steps to writing my own book.

4 years after writing the chapter, some elements needed to be reviewed, but the future vision remained relatively stable. Because it resonates so well with most of the messages of this book, I decided to keep this as the closing chapter to give you a scenario of how the world, and more specifically, the financial services industry could look like if they define a new Ethos, embracing ethics and values for a meaningful transformation.

This chapter is an updated version of the one that I originally wrote in 2020 for the book "Digital for Good - Stand for something or you will fall" by Chris Skinner. The book was originally published in 2021.

A final thought for the future, and the story of Leon in 2040.

Leon never considered changing his bank, the same bank as his parents chose at his birth, why should he? Now, it starts to annoy him that an institution selected by his parents holds his most trusted assets. So, Leon recently started looking at the options for a more independent life. Why is he looking?

R. Coeckelbergs, *A New Ethos in Banking*,
https://doi.org/10.1007/978-3-031-78234-3_10

First of all: the bank that his parents chose merged a few times so, in a way, that bank was no longer the same bank. After 20 years of consolidation in the industry, the bank no longer stood for the same values it used to represent 20 years ago.

Second, there is no, and there has never been, any emotional bonding with the bank that holds most of his assets.

Thus far, Leon held a passive link with his bank. The trusted relationship was just to keep his digital assets, his digital identity, safe. It provided every basic need that one would expect from a bank in 2040 … expectations that were entirely different from those when he was born back in 2020.

He doesn't really feel connected with that bank now. Leon grew up and he changed over the years, as the bank did. Now he feels that his bank no longer matches his needs or values. For the first time, it annoys him that he does not feel connected.

10.2 COVID-19 Changed People's Relationship with Owning "Stuff" and Money

Banking was not on top of his list of interests until now because, as long as Leon remembers, his dad worked for the financial services industry. His dad had told him all of these stories of how the industry changed over the years. He listened to the shifts that happened and how banking completely transformed. He had to listen as his dad always tried to frame the transformations in a "child-friendly" manner, whatever that means. His dad's view was that if my kid understands it, everyone will.

"The complexity", his dad told him, "is not so much which bank you want to hold your financial relationship with, the question is which values you want to live? What life do you want? Any bank can fulfil your financial needs, the question is: how they do this, based on which fundamentals? What are the bank's aspirations to serve your financial needs? Your bank is supposed to be a life partner, someone you can trust your life with. Do you trust your life with your bank?"

Twenty years ago, Europe had over 6,000 banks, all offering the same products and services. People picked the one they liked—often the cheapest one or because they had a branch nearby—and switched when they needed a mortgage or similar large loan, as this had the biggest and the most visible cost attached. The rise of the internet changed the way people banked, but it didn't change much where and how people banked.

The global pandemic in 2020–2021 changed this trend. What started as a global health crisis ended up being the turbo charge accelerator of digital transformation. The dot-com bubble was characterised by an excess of supply in digital solutions, but the demand was not there. COVID-19 created it. Consumers adopted digital solutions and quickly started cherry-picking the digital providers with the best customer experience. Convenience was no longer enough to keep customers happy. Making traditional processes available through digital channels did no longer suffice. Consumers expected more.

Leon had heard lots of ideas about concepts like AI, big data and data modelling, peer-to-peer and instant payments, from the stories his dad told him. Technology and software supporting these trends have obviously completely democratised today, but back then?

According to Leon's dad, the notion of digital transformation remained a hot debate until the mid-2020s. Around 2030, this changed as expectations from the market rebalanced. A fit-for-purpose digital architecture, digital platforms, digital service offerings, and production were a prerequisite. Ten years after the debate on ESG became a mainstream discussion that affected everyone, mainstream banks finally adopted the concept of purposeful banking and banking for good as part of their being instead of another list of regulations.

What counts today, in 2040, is how banks employ it to fulfil their purpose for society and the people they serve.

The 2020 pandemic also changed banks' focus drastically. Interest rates hiked a bit to regain control over inflation, but remained low for a long time, which squeezed the profits. On top of that, consumers lost interest in credit services, for a couple of reasons. By 2027,[1] indebtedness will become a disruptive concern, not just for consumers but also for businesses and countries alike. Although the economy recovered from this, the mindset of the people towards credit changed. They started to understand the impact of a credit that cannot be paid back.

First of all, credit lines, buy-now-pay-later and similar products came under scrutiny. In the aftermath of the crisis, many consumers fell back on short-term credit, which became a gallows for society's weakest. Politics and regulation intervened, making these products a lot less appealing for credit institutions, alternative payment providers and banks.

On top of that, and maybe even more important than the more stringent regulatory framework, were macro-economic circumstances. The circular economy shifted the way people thought about ownership, which reduced the need for credits to "buy stuff".

Usage-based upon "everything-as-a-service" was the way forward. People stopped buying things and just paid for using things, resulting in a higher quality of products and more balanced consumption patterns by consumers using it. Since they had more flexibility to switch, people used what they could afford instead of buying something based on future returns, for which they needed credit in the meantime.

By 2030, people were consuming more than ever before, instead of purely buying things, and banks were the helpful partners for consumers to make sure they got the right products and services at the right price.

The evolution from open banking to open finance and open data unlocked unexpected opportunities. On top of integrating their services in all kinds of customer journeys, banks created new services. By 2040, cultural elements dominated the way in which banks worked, rather than technology. For example, partnership banking and Islamic banking became a big thing in the circular economy, where banks allowed people to invest in community projects at their own pace.

More on that later.

10.3 Contextual Integration and the End of Daily Banking

In the early 2020s, lots of start-up firms, all thrown in a bucket called "fintechs", democratised payments, and daily banking got a lot less attractive for banks. Still, they realised that they needed to offer this service in order to strengthen their relationship with their customers. Leon's father told him that adding loyalty products helped for a few years, but other companies quickly took over the same principles with better deals and services. Banks needed to offer more than a discount on impulse purchases to remain relevant in this space.

By the late 2020s, financial institutions finally understood the power of open finance. After years of discussing the topic, and one regulation and directive after the other (PSD2, PSD3, PSR, FIDA) the fintech industry's movement finally paved the way to the required standardisation to scale open finance opportunities to something that could be global.

As a result, many banks were pushed behind the scenes, and the idea of banks becoming dumb pipes without any connection with the end customer ended up being an unfortunate reality for many in the context of transaction processing. Most of these "dumb pipes" fell victim to the massive consolidation in the banking industry.

As Leon's dad would say: "In the past, before the internet, banks were relevant for the physical footprint. With the internet, the world entered a global village, and distance was no longer a concern. Suddenly these banks realised that they all looked the same". Banks missed out on clearly explaining their unique purpose; their reason for being.

Out of this consolidation wave, a few big players took the opportunity for open finance to be omnipresent. They integrated into every possible relevant context of everyone's lives, always available, whenever needed. They'd pop up to serve their customers, and they popped up the way their customers liked it. Some were centuries-old banks, but not all. In fact, a few of these dominant financial services providers have only been around for a decade or two and have quickly evolved into fully regulated financial institutions.

In the same period, Central Bank Digital Currencies reached a critical mass, and fintech companies that survived the consolidation led the innovation efforts in payments and the management of day-to-day assets. Payments and current accounts were no longer called "daily banking" because they no longer fit a bank's purpose. Banks were simply pushed out of the equation in the payments space.

Today, in 2040, payments are no longer the sticky product it used to be, but banks kept a finger on the pulse through the cooperative Swift. Swift is a network that first only serviced its founding members, the banks and later also big corporates. After transforming their messaging business to a platform business, they reached out to fintechs for more value-added services, and by 2025 they started to attract SME customers.

Ten years later, by 2035, Swift was the platform that connected all banks, where consumers used a unique ID, with multiple accounts at different providers. As a trusted partner, banks took care of identity management and a digital asset custodian for non-day-to-day usage. Oftentimes banks were the ones that manage the oversight across different fintechs and other wallets.

"With some delay", Leon's dad explained, "banks were living a similar scenario to retailers. They witnessed an evolution that started around 2000 and accelerated in 2010–2020 where a few big players grew bigger by connecting with all kinds of alternative players to increase their market share". It was all down to technology allowing total transparency, allowing banks to improve their services, allowing privacy whenever needed, and a regulator to avoid abuse of the market concentration.

Now, Leon owns physical cash, but he buys it for fun, as a collectable, like his grandparents' stamp collection, as it can no longer be used to make payments. He got it as a gift from his father at the age of 18, but it is no longer a generally accepted means of exchange as all of his payments are made digitally.

He earns a few bucks with jobs he finds online whenever he is desperate for money or has some spare time. Occasionally, he opts to work for premium delivery services. It keeps him fit with the bike, and he also then gets paid for it!

According to his dad, this is not much different from 20 years ago, when they called people like him a gig worker, except for the fact that, back then, the best earners were the fastest delivery people. Today, premium delivery is more therapy and listening to customers than anything else. Free, instant delivery is all drone-based. Most of the time, though, he prefers more creative projects or gaming, which he would do from home or wherever he is.

Leon's dad tells him that what is different today is that each platform issues its own tokens. In fact, platforms are issuing their own currencies in this context. Tokens are currencies. Most of these tokens can be used anywhere, but if you use them for services on that platform, you get a discount. It's the latest loyalty scheme. Interestingly, some of these platforms added a specific tag in their currency to limit the possibilities to buy things so, for example, you could only buy sustainable goods.

Whenever he reaches a particular surplus of tokens, the money is automatically moved to his savings wallet to ensure that he doesn't spend all his excess savings. All of this is automatically managed by his bank.

This is not a superfluous luxury since payments are pushed to the back. In fact, by 2030, payments got truly seamless and invisible through a mix of biometrics, location-based technology and device detection. Payments got so convenient that the regulator needed to step in to make sure "the pain of paying" was re-introduced to the process in an effort to reduce financial stress in society. That is why now, in 2040, you need to explicitly consent or be aware of the trigger for a purchase that results in a payment. You get a notification upfront in case of insufficient funds and a request to pay with your main currency or an alternative currency from another issuer.

Leon has complete oversight over the assets he holds, while the aggregated data gives him a complete overview of his assets under management, as well as earning money on the side from anonymised insights, an alternative revenue stream through the open data movement. This keeps him financially well, he learns how to get financially better, and it provides him with more savings.

He holds small amounts of coins at all these platforms, but he has only one savings bank. His savings bank, a piggy bank if you like, is a wallet solution where he saves whatever he cannot miss, both digital currencies or any other digital asset. That wallet is safe at his bank … the bank he is now reconsidering.

10.4 Responsible Banking

The industry understood that educating people was a lot harder to realise than simply offering solutions to take away the pain and the need for financial education. Financial education was a hot topic ten years ago, but today, it is a lot less important.

Financial education courses were demoted after a wave of accelerated promotion when governments viewed financial education as *a life skill*. Then, technological advancements made financial products and services completely complicated. Technology created a lot of convenience, but the customers remained responsible for their decisions, as complex as financial services were. What followed was technology that reached the level that it was so reliable that banks took care of helping their customers with the best possible solution for their unique persona. Everything was individualised.

The digital advisor would automatically select the best possible way to pay or save. Banks proactively made clients healthier. Like a doctor who tells the client which medicine to take, the bank would give the customer a choice of conscious or unconscious banking.

People who didn't want to bother consciously dealing with their tokens or their money didn't have to (except for the consent at the moment of purchase). Banks took care of it, wholly integrated into a consumer's day-to-day life. Most people chose unconscious banking, at least in their daily banking activities. Banking became invisible.

In doing so, banks began to act as being like doctors and therapists, since reduced financial stress resulted in lower mental stress and better health in general. People are no longer purely trusting their money with the bank. Instead, they have started to trust their life with them to some extent. Not only did it increase the well-being of people, but it also improved the cost of social welfare and insurance.

By way of example, customers can now (in 2040) choose whether they could use gambling, alcohol or any other addictive services or not. Equally, as part of the client relationship, banks would detect sensitivity for problematic gambling or alcohol usage and would act upon it by alerting the customer and, if given permission, the authorities.

By the 2030s, banks understood that they were enablers for a better future. Instead of just selling products, banks finally understood that their impact, relevance, and purpose were far beyond that. Responsible banking became a lot more than what the initials ESG implied. Without the right stakeholder management, banks could no longer survive. Green finance, social, and societal responsibilities became the fundamental basics for banking, no longer a diversifying element. More than this, banks finally understood what it means to focus on customer journeys, across industries, rather than just banking.

10.5 The Rise of Partnership Banks

Besides the few global institutions that built excellent universal products at a low cost, community banks and savings banks remained relevant and essential players because of life event management. They focused far more on the customer's lives and life cycle, than just the transaction or payment. With the right measures and synergies, they improved their cost structure, simplified their internal processes and anchored themselves in the local economy.

They built their business around their purpose, their unique corporate identity and their link with the community. Proximity became their main differentiator, and, by that, it was not so much their physical proximity per se but their close ties to the local community, its citizens and particular niches of customers. These were mostly cooperative initiatives, with lower profits but more ownership and a reciprocal relationship with their customers and society.

Community banks would have a community centre, with the ability to meet clients to look for financial advice, and with the added advantage of empathetic and human remote advice to keep this person-to-person contact to a minimum. The rest of the time, these community centres held all kinds of local activities that governments could no longer afford to organise, such as acting as a lesson centre for flower arrangements, childcare, theatre, co-working, and more. These centres glued the community together, adding to the social role that banks appropriated themselves.

This reciprocal relationship put "responsible banking" at the forefront of the agenda. These niche banks took "responsible banking" to the level of creating more societal cohesion as a trusted partner to local governments, enterprises, associations, and inhabitants. Whether it was the national dentist association, a university or Greenpeace, they all had their own bank. They did not take care of payments, but provided a trusted financial and identity service to assist their customers.

Depending on where these niche players put their emphasis, they all built their business models on the principles of open finance and proprietary finance solutions. Leon's dad is probably right in saying that, technically speaking, most of these are not a bank, but just a brand supported by banks, but people saw them as banks.

Today, banks support society with several initiatives for a cleaner, healthier, more peaceful, and better world. They fully play the role of the financial intermediary to make sure that money is used to create a better world.

10.6 Purposeful Money

Financial intermediation is one of the traditional bank's core activities: making sure that excess deposits are invested back into the economy through lending and investments. Deposits are aggregated and transformed into credits. Part of the profits go to the deposit account (the interest), and part is a risk premium for the bank (the exposure). The latter will cover potential losses, so the deposit account holder stays out of that equation.

Since the 2020s pandemic, debt remained a concern for a considerable part of the population. It was not so much because they couldn't afford it, but because they perceived it as a risk for their future. As a consequence, banks started picking some of the ideas of Islamic banking.

Modern Islamic banking started with Islamic businesspeople who studied traditional banking and created a religious, ethical version of it. The traditional banks then took on board some of the principles of these Islamic banks. Most of them did not become Sharia-compliant, but the concept of partnership banking was a success. The aim was to create financial products that were independent of interest and focused far more on the asset, the thing that was being bought.

Banks focused more on investments instead of issuing loans. Customers who liked to invest some of their assets had a valuable partner with these banks. The customer explained what he wanted to invest in, and the bank would buy the asset from him. Instead of paying off debt over time, the customer would buy small parts of ownership.

Banks made grateful use of the data revolution and strengthened their risk management skills. A new business was born based on purpose-driven ethical banking. Instead of measuring the risk of a customer, banks measured the risk of the asset for which a customer required financial support. Partnership banking meant the risks and profits were shared with their customer. It reflected the rise of stakeholder capitalism, where all stakeholders need to be served and not just the shareholder.

Depending on the client, the bank would look for the best possible solution. Customers with a good credit rating got the choice to either take a regular loan, with immediate ownership of the assets or work with the bank as a partner in the project. A starting dentist was no longer dependent on a heavy loan to start a new practice, as they fell back on a renting agreement with the specialised bank.

It stimulated financial inclusion, and it made life better for all the people who struggled with credit. It completely transformed how the world looked at financial intermediation, and it speeded up the idea of Products-as-a-Service, with banks in the driver's seat.

10.7 Tokenisation of Everything

In the '20s people were saying that, thanks to tokenisation, we would no longer need banks. The idea was that tokenising everything together with crowdfunding would disintermediate banks from the economy.

Tokenisation did have a significant impact on banking. Aside from digital currencies, the tokenisation of everything speeded up the revolution of partnership banking. It created transparency in the valuation of all kinds of assets, resulting in an immense inclusion of people within communities.

Housing-as-a-Service became a thing and, with every month someone paid their rent, they would own a small percentage of the apartment. The same is true for all the other *as-a-service* in a community. People could start building up ownership at their own pace, and banks supported them through flexible partnership banking models.

The global banks took care of more significant projects while smaller, local community banks, thrived in this circular economy, as it brought banking a lot closer to the linked community or association. It allowed people to have a voice in how the assets were managed.

10.8 What Are You Looking for?

"What are you looking for?" Leon's dad asks. That sounds a lot easier than it is. Leon just wants a bank that helps him make his life more comfortable and grow a financial buffer for the future (some things never change).
"I want a bank that thinks with me", Leon replies.
"Why don't you stick to your current bank, give it a few more years and see where you want to go with your life?"

"What for?" Leon asked. "Of course, community banks are very much tied to the local community, and who knows where I'll live in a few years, but this should not stop me from thinking hard about which bank to trust with my assets and my financial health, right?"
"You're right. Maybe you could consider one of the niche players for the student community? Their services are built for people like you, and they may even guide you to another institution from the moment you no longer fit in their customer profile?" his dad answers.
"Community banks are no longer purely linked to geographic communities. Today, you can have a Game of Thrones bank, Tottenham Hotspur bank, Dentist Association bank, or any other bank that represents your interests, including the local university community bank".
"These banks take their role seriously. Most of these banks work with one of the few global banks, but they do an excellent job of servicing their young clients and preparing them for adult life. They provide well-documented suggestions on which banks to partner with next".

What is so special about them are partnerships with different brands that give students great deals on all kinds of products and services. More important is the way that they organise student loans. Instead of simply providing a loan that can be paid back when someone starts working, they actively look for solutions to reduce the liabilities to the bank as soon as possible.

These banks actively monitor the student's earnings and spending pattern. They advise their young clients with the ultimate goal of paying back the student loan earlier than agreed in the initial contract, without additional charges.

Although Leon does not have any student loans right now, his parents saved a required amount over the years to make sure his student years did not make him financially dependent before starting adult life, and he gets the point his dad made.

So, he applies for a bank ID at his University community bank.

Note

1. In the original text, in 2020 I estimated this to be 2025.

Bibliography

ABN AMRO Bank N.V. (2023). *ABN AMRO passes mark of 100 Financial Care Coaches*. Opened on July 1, 2024, available at https://www.abnamro.com/en/news/abn-amro-passes-mark-of-100-financial-care-coaches

Adviescommissie Toekomst Banken. (2009). *Naar herstel van vertrouwen*. Opened on March 30, 2024, available at https://www.nvb.nl/media/1668/000603_000593-rapport-adviescommissie-toekomst-banken-def.pdf

Advisory Board. (2023). *What ChatGPT does better than doctors, according to one study*. Opened on July 15, 2024, available at https://www.advisory.com/daily-briefing/2023/05/02/chatgpt-empathy

Alessa.com. (2023). *The money laundering control act of 1986: An overview for AML compliance professionals*. Opened on November 11, 2023, available at https://alessa.com/blog/money-laundering-control-act/

Alikhan, S. (2024). *Silicon Valley Bank collapse one year on: What was the impact?* Opened on July 3, 2024, available at https://www.finextra.com/the-long-read/970/silicon-valley-bank-collapse-one-year-on-what-was-the-impact

Birch, D. (2017a). Before Babylon, Beyond Bitcoin—From money that we understand to money that understands us. In D. Birch, *Before Babylon, Beyond Bitcoin—From money that we understand to money that understands us* (pp. 5–7). London Publishing Partnership.

Birch, D. (2017b). Before Babylon, Beyond Bitcoin—From money that we understand to money that understands us. In D. Birch, *Before Babylon, Beyond Bitcoin—From money that we understand to money that understands us* (p. 8). London Publishing Partnership.

Bollough, O. (2018). *Moneyland—Why thieves & crooks now rule the world & how to take it back* (pp. 38–41). Profile Books LTD.

R. Coeckelbergs, *A New Ethos in Banking*,
https://doi.org/10.1007/978-3-031-78234-3

Broløs, A., & Taylor, B. E. (2024a). Designing for people—Learning from inclusion initiatives. In *Customer-centric innovation in finance—Leveraging human insights to drive product innovation in the digital age.* Kogan Page.

Broløs, A., & Taylor, E. (2024b). In *Customer-Centric Innovation in Finance: Leveraging human insights to drive product innovation in the digital age* (pp. 231–233). Kogan Page.

Bullough, O. (2019a). *Moneyland—Why thieves & crooks now rule the world & how to take it back* (pp. 286–288). Profile Book Ltd.

Bullough, O. (2019b). Moneyland—Why thieves and crooks now rule the world and how to take it back. In O. Bullough, *Moneyland—Why thieves and crooks now rule the world and how to take it back.* Profile Book Ltd.

Cambridge Dictionary. (2023). *Cambridge Dictionary—Ethos.* Opened on December 10, 2023, available at https://dictionary.cambridge.org/dictionary/english/ethos

Candelon, F., Krayer, L., Rajendran, S., & Martinez, D. Z. (2023). *How people can create—And destroy—Value with generative AI*. Opened on July 15, 2024, available at https://www.bcg.com/publications/2023/how-people-create-and-destroy-value-with-gen-ai

Carbon Dislosure Project. (2017). *New report shows just 100 companies are source of over 70% of emissions.* Opened on September 5, 2024, available at https://www.cdp.net/en/articles/media/new-report-shows-just-100-companies-are-source-of-over-70-of-emissions

Carbo-Velverde, S. J., Cuadros-Solas, P., Gonnella, C., & Rodriguez-Fernandez, F. (2023). The digtialization of the European banking industry: Some evidence. In P. Molyneux (Red.), *New challenges for the banking industry—Searching for balance between corporate governance, sustainability and innovation* (pp. 277–278). Palgrave Macmillan Studies in Banking and Financial Institutions.

Caudron, J. (2023). De Toekomstformule. Pelckmans.

CFPB. (2023). *CFPB proposes rule to jumpstart competition and accelerate shift to open banking*. Opened on May 5, 2024, available at https://www.consumerfinance.gov/about-us/newsroom/cfpb-proposes-rule-to-jumpstart-competition-and-accelerate-shift-to-open-banking/

Climate Policy Initiative. (2023, November 30). *Global landscape of climate finance 2023.* Opened on August 8, 2024, available at https://www.climatepolicyinitiative.org/wp-content/uploads/2023/11/Global-Landscape-of-Climate-Finance-2023.pdf

Coeckelbergs, R. (2019a, April 19). *How to find great collaboration between fintech and banks.* Opened on January 7, 2024, available at https://thebankingscene.com/opinions/how-to-find-great-collaboration-between-fintech-and-banks

Coeckelbergs, R. (2019b, July 15). *The Open Banking Interviews: Karin Van Hoecke, General Manager Digital Transformation, KBC Bank & Insurance Belgium.* Opened on December 15, 2023, available at https://thebankingscene.com/opinions/obi-karin-van-hoecke-general-manager-digital-transformation-kbc-bank-insurance-belgium

Coeckelbergs, R. (2020). *Islamic banking lessons and the future of Islamic Fintech.* Opened on August 15, 2024, available at https://thebankingscene.com/opinions/islamic-banking-lessons-and-the-future-of-islamic-fintech/

Coeckelbergs, R. (2021). *Banking for good—A vision by vdk bank.* Opened on August 5, 2024, available at https://thebankingscene.com/opinions/banking-for-good-a-vision-by-vdk-bank/

Coeckelbergs, R. (2022a). *ABN AMRO Bank's sales transformation with self-organising teams.* Opened on July 23, 2024, available at https://thebankingscene.com/opinions/abn-amro-banks-sales-transformation-with-self-organising-teams/

Coeckelbergs, R. (2022b). *Rabobank's global food revolution.* Opened on August 5, 2024, available at https://thebankingscene.com/opinions/rabobanks-global-food-revolution/

Coeckelbergs, R. (2022c). *The adventures of Toddler KATE, KBC's virtual assistant.* Opened on August 5, 2024, available at https://thebankingscene.com/opinions/the-adventures-of-toddler-kate-kbcs-virtual-assistant/

Coeckelbergs, R. (2022d, June 8). *Digital inclusion—Making digital banking available for all.* Opened on January 7, 2023, available at https://thebankingscene.com/opinions/digital-inclusion-making-digital-banking-available-for-all

Coeckelbergs, R. (2022e, October 3). *Digital inclusion in Belgium Anno 2022.* Opened on January 7, 2024, available at https://thebankingscene.com/opinions/digital-inclusion-in-belgium-anno-2022

Coeckelbergs, R. (2022f, October 3). *Digital inclusion in Belgium Anno 2022.* Opened on January 7, 2024, available at https://thebankingscene.com/opinions/digital-inclusion-in-belgium-anno-2022

Coeckelbergs, R. (2023a). *A new era in banking: How did we get here? An interview with Niccolo Polli—Head of strategy and sustainability at HSBC Europe.* Opened on January 10, 2024, available at https://thebankingscene.com/opinions/a-new-era-in-banking-how-did-we-get-here-an-interview-with-niccolo-polli-head-of-strategy-and-sustainability-at-hsbc-europe/

Coeckelbergs, R. (2023b). *The Greek credit tragedy and the new ethos in banking.* Opened on March 02, 2024, available at https://thebankingscene.com/opinions/the-greek-credit-tragedy-and-the-new-ethos-in-banking

Coeckelbergs, R. (2023c, February 28). *Leveraging finance for social impact—Perspectives by Erste group.* Opened on January 4, 2024, available at https://thebankingscene.com/opinions/leveraging-finance-for-social-impact-perspectives-by-erste-group/

Coeckelbergs, R. (2023d, November 28). *When art and tech dance together—A review on augmented empathy talks.* Opened on March 14, 2024, available at https://thebankingscene.com/opinions/when-art-and-tech-dance-together-a-review-on-augmented-empathy-talks

Coeckelbergs, R. (2023e, November 6). *Setting the scene of ethos in banking.* Opened on November 7, 2023, available at https://thebankingscene.com/opinions/setting-the-scene-of-ethos-in-banking

Coeckelbergs, R. (2023f, December 14). *About David, Goliath and The New Ethos in Banking*. Opened on January 7, 2024, available at https://thebankingscene.com/opinions/about-david-goliath-new-ethos-in-banking-credit-union-mutual

Coeckelbergs, R. (2023g, December 4). *About Mary Poppins and the New Ethos in Banking*. Opened on December 13, 2023, available at https://thebankingscene.com/opinions/about-mary-poppins-and-the-new-ethos-in-banking-interview-michael-anseeuw-ceo-bnp-paribas-fortis

Coeckelbergs, R. (2024a). *How regulation is affecting the rules of engagement in banking*. Opened on February 12, 2024, available at https://thebankingscene.com/opinions/how-regulation-is-affecting-the-rules-of-engagement-in-banking

Coeckelbergs, R. (2024b). *How regulation is affecting the rules of engagement in banking*. Opened on February 14, 2024, available at https://thebankingscene.com/opinions/how-regulation-is-affecting-the-rules-of-engagement-in-banking

Coeckelbergs, R. (2024c). *How the sustainable agenda affects the rules of engagement between banks and their clients*. Opened on March 4, 2024, available at https://thebankingscene.com/opinions/how-the-sustainable-agenda-affects-the-rules-of-engagement-between-banks-and-their-clients

Coeckelbergs, R. (2024d). *The banker's oath and the new ethos in banking (part 1 of 2)*. Opened on March 30, 2024, available at https://thebankingscene.com/opinions/the-bankers-oath-and-the-new-ethos-in-banking-part-1-of-2

Coeckelbergs, R. (2024e). *The banker's oath and the new ethos in banking (part 2 of 2)*. Opened on March 31, 2024, available at https://thebankingscene.com/opinions/the-bankers-oath-and-the-new-ethos-in-banking-part-2-of-2

Coeckelbergs, R. (2024f). *The future role of banks in our society—Interview with Guy Schellinck, CEO, Beobank*. Opened on August 5, 2024, available at https://thebankingscene.com/opinions/the-future-role-of-banks-in-our-society-interview-with-guy-schellinck-ceo-beobank/

Coeckelbergs, R. (2024g). *The new ethos in banking: Values based banking—Interview with Martin Rohner*. Opened on September 10, 2024, available at https://thebankingscene.com/opinions/the-new-ethos-in-banking-values-based-banking-interview-with-martin-rohner/

Coeckelbergs, R. (2024h). *The triple-sided coin and the new ethos in banking*. Opened on August 5, 2024, available at https://thebankingscene.com/opinions/vdk-bank-triple-sided-coin-and-the-new-ethos-in-banking/

Coeckelbergs, R. (2024i). *We conclude our talk with Ethicist Joris Krijger about The New Ethos in Banking (Part 2 of 2)*. Opened on September 5, 2024, available at https://thebankingscene.com/opinions/we-conclude-our-talk-with-ethicist-joris-krijger-about-the-new-ethos-in-banking-part-2-of-2/

Coeckelbergs, R. (2024j, January 19). *The Holy Trinity in customer centricity*. Opened on January 21, 2024, available at https://thebankingscene.com/opinions/the-holy-trinity-in-customer-centricity-interview-stefaan-rogiers-bil

Coeckelbergs, R. (2024k, February 5). *Unveiling the future of finance—A comprehensive exploration of embedded finance perspectives*. Opened on August 30, 2024, available at https://thebankingscene.com/resources/unveiling-the-future-of-finance-2/

Consumer Financial Protection Bureau. (2016, September 8). *Consumer financial protection bureau fines wells Fargo $100 million for widespread illegal practice of secretly opening unauthorized accounts.* Opened on December 14, 2023, available at https://www.consumerfinance.gov/about-us/newsroom/consumer-financial-protection-bureau-fines-wells-fargo-100-million-widespread-illegal-practice-secretly-opening-unauthorized-accounts/

Consumer Financial Protection Bureau. (2022, December 20). *Consumer financial protection bureau CFPB orders wells Fargo to pay $3.7 billion for widespread mismanagement of auto loans, mortgages, and deposit accounts.* Opened on December 14, 2023, available at https://www.consumerfinance.gov/about-us/newsroom/cfpb-orders-wells-fargo-to-pay-37-billion-for-widespread-mismanagement-of-auto-loans-mortgages-and-deposit-accounts/

Copeland, R. (2024). *One year after bank crisis, a struggle over what needs to change.* Opened on July 1, 2024, available at https://www.nytimes.com/2024/03/10/business/silicon-valley-bank-one-year-regulations.html

Cranston, S., & Keller, S. (2013). *Increasing the 'meaning quotient' of work.* Opened on August 02, 2024, available at https://www.mckinsey.com/capabilities/people-and-organizational-performance/our-insights/increasing-the-meaning-quotient-of-work

Cristina, R. (2022). Ethics in banking—Is it possible? In C. Rovera, *Ethics in banking—Is it possible?* (p. 24). Springer Nature Switzerland.

Danske Bank. (2019). *Danske Bank to compensate Flexinvest Fri customers.* Opened on July 3, 2024, available at https://danskebank.com/-/media/danske-bank-com/pdf/news-insights/company-announcements/2019/uk/danske-bank-press-release-24-june-2019.pdf?rev=32fa72b4e5504ae48cad534e9ab2e62c&hash=D34464649323375F2402F9DD6DFA1A27

Danske Bank. (2021). *Carsten Egeriis replaces Chris Vogelzang as CEO of Danske Bank A/S. Gerrit Zalm resigns from the Board of Directors.* Opened on July 3, 2024, available at https://danskebank.com/news-and-insights/news-archive/company-announcements/2021/ca19042021

Danske Bank. (2022). *Danske Bank reaches coordinated resolutions with the US and Danish authorities regarding the Estonia matter.* Opened on July 3, 2024, available at https://danskebank.com/news-and-insights/news-archive/company-announcements/2022/ca13122022b

Dankse Bank Group. (2024). *Climate Action Plan Progress Report 2023.* Opened on March 10, 2024, available at https://danskebank.com/-/media/danske-bank-com/file-cloud/2024/2/climate-action-plan-progress-report-2023.pdf?rev=8e354015e7374d3b90f3fcc4ae83da78&hash=A0683DB98DA133E04A5B2A8FA6488E9B

Danske Bank. (2024a). *Dankse Bank—Our purpose.* Opened on April 15, 2024, available at https://danskebank.com/about-us/our-purpose

Danske Bank. (2024b). *The bank moves out, farming moves in 1871–1910*. Opened on March 10, 2024, available at https://danskebank.com/about-us/our-history/the-bank-moves-out-farming-moves-in-1871-1910

Da Rin, M., & Helman, T. (2002, August). Banks as Catalysts for Industrialization. *Financial Markets Group, The London School of Economics and Political Science*, 4–11.

De Vlies, P. (2016). *Bankgeheim* (p. 92). Uitgeverij Lannoo Campus.

Dinesen, C. (2020a). Absent management in banking—How banks fail and cause financial crisis. In C. Dinesen (Ed.), *Absent management in banking—How banks fail and cause financial crisis* (pp. 16–31). Palgrave Macmillan.

Dinesen, C. (2020b). Absent management in banking—How banks fail and cause financial crisis. In C. Dinesen, *Absent management in banking—How banks fail and cause financial crisis* (p. 47). Palgrave Macmillan.

Dinesen, C. (2020c). Absent management in banking—How banks fail and cause financial crisis. In C. DInesen, *Absent management in banking—How banks fail and cause financial crisis* (pp. 51–175). Palgrave Macmillan.

European Banking Authority. (2015, July 15). *Guidelines on product oversight and governance arrangements for retail banking products*. Opened on July 9, 2024, available on https://www.eba.europa.eu/sites/default/files/documents/10180/1141044/d84c9682-4f0b-493a-af45-acbb79c75bfa/EBA-GL-2015-18%20Final%20report%20on%20Guidelines%20on%20product%20oversight%20and%20governance.pdf

European Commission. (2011, October 25). *Communication from the commission to the European Parliament, the council, the European economic and social committee and the committee of the regions a renewed EU strategy 2011–14 for Corporate Social Responsibility*. Opened on January 7, 2024, available at https://eur-lex.europa.eu/legal-content/EN/TXT/?uri=CELEX:52011DC0681

European Commission. (2023). *Modernising payment services and opening financial services data: New opportunities for consumers and businesses*. Opened on July 1, 2024, available at https://ec.europa.eu/commission/presscorner/detail/en/ip_23_3543

European Parlaiment. (2019). *A decade on from the crisis—Main responses and remaining challenges*. Opened on March 17, 2024, available at https://www.europarl.europa.eu/RegData/etudes/BRIE/2019/642253/EPRS_BRI%282019%29642253_EN.pdf

European Payments Council. (sd). *SEPA Payment Account Access*. Opened on July 1, 2024, available at https://www.europeanpaymentscouncil.eu/what-we-do/other-schemes/sepa-payment-account-access

EY & ABBL. (2021). *Survey on the cost of regulation and its impact on the Luxembourg financial centre*. Opened on July 9, 2024, available at https://assets.ey.com/content/dam/ey-sites/ey-com/en_lu/topics/banking-and-capital-markets/ey-luxembourg-survey-on-the-cost-of-regulation-2021-print.pdf

Eynikel, J. (2021a). In *Check in—The search for meaning and purpose in business* (pp. 123–128; 137–147). Lanno.

Eynikel, J. (2021b). In *Check-in: The search for meaning and purpose in business* (p. 143). Lannoo Campus.

Eynikel, J. (2021c). The meaning of meaning. In *Check in—The search for meaning and purpose in business* (pp. 104–105). Lannoo.

Febelfin. (2023, June 29). *Digital payments, already well-rooted in the habits of Belgians, continue to increase*. Opened on March 14, 204, available at https://febelfin.be/en/press-room/digitalization-innovation/digitaal-betalen-stevig-verankerd-in-gewoonten-van-belg-en-blijft-groeien

Febelfin. (2024). *Banker's oath is introduced*. Opened on March 30, 2024, available at https://febelfin.be/en/press-room/supervision-policy/bankers-oath-is-introduced

Federal Deposit Insurance Corporation. (2023). *The savings and loan crisis and its relationship to banking*. Opened on June 3, 2024, available at https://www.fdic.gov/resources/publications/history-eighties/volume-1/history-80s-volume-1-part1-04.pdf

GABV. (2023). *vdk bank created first ever three-sided coin*. Opened on August 8, 2024, available at https://www.gabv.org/think-bolder/vdk-bank-creates-first-ever-three-sided-coin/

Gartenberg, C., Prat, A., & Serafeim, G. (2019, January 19). *Corporate purpose and financial performance*. Opened on August 5, 2024, available at https://www.ecgi.global/sites/default/files/Corporate%20Purpose%20and%20Financial%20Performance-%20Paper.pdf

Girling, W. (2021). *The creation and evolution of credit card payments*. Opened on July 1, 2024, available at https://fintechmagazine.com/digital-payments/creation-and-evolution-credit-card-payments

Goclowski, M. (2019). *Reuters—Polish banks haunted by Swiss franc mortgages in European court*. Opened on November 10, 2023, available at https://www.reuters.com/article/us-poland-banks-swissfranc-analysis-idUSKBN1WE044

Harper, J. (2023). *DW.com—What's at stake in Poland's Swiss franc case?* Opened on November 30, 2023, available at https://www.dw.com/en/whats-at-stake-in-polands-swiss-franc-case/a-65920810

Heimbach, D. R. (2015). *How the term "ethics" has evolved*. Opened on July 1, 2024, available at https://erlc.com/research/how-the-term-ethics-has-evolved/

Hendrickson, J. M. (2011). In *Regulation and instability in U.S. commercial banking—A history of crises*. Palgrave Macmillan.

Hinssen, P. (2024). *AI and the End of Awful*. Opened on July 27, 2024, available at https://www.peterhinssen.com/assets/files/books/AI-and-the-End-of-Awful-e-book.pdf

International Monetary Fund. (2023). *The IMF and the fight against money laundering and terrorism financing*. Opened on July 1, 2024, available at https://www.imf.org/en/About/Factsheets/Sheets/2023/Fight-against-money-laundering-and-terrorism-financing

International Renewable Energy Agency. (2023). *World Energy Transitions Outlook 2023 VOLUME 1.* Opened on August 5, 2024, available at https://www.irena.org/Digital-Report/World-Energy-Transitions-Outlook-2023

Kanaal, Z. (2024). *Digitale assistent Kate bespaart KBC 200 werknemers in België.* Opened on August 8, 2024, available at https://trends.knack.be/kanaal-z/z-nieuws/bekijk-digitale-assistent-kate-bespaart-kbc-200-werknemers-in-belgie/

Karamichaildou, G., Margaritis, D., & Mayes, D. G. (2017). Assysemmetry, austerity and anxiety: The approach to the Greek debt crisis. In *The Greek debt crisis—In quest of growth in times of austerity* (pp. 45, 69–71). Springer Nature.

KBC Bank. (2024). *De KBC-Belmobiel komt naar je toe*! Opened on July 1, 2024, available at https://www.kbc.be/particulieren/nl/contact/kbc-belmobiel.html

Kerkez, N. (2023). *International Banking—The cost of compliance—When will investment in risk technology start paying off for banks*? Opened on July 9, 2024, available at https://internationalbanker.com/finance/the-costs-of-compliance-when-will-investment-in-risk-technology-start-paying-off-for-banks/

Khalaf, R. (2019). *Financial Times—Danish regulator reports Danske to police over mis-selling.* Opened on March 10, 2404, available at https://www.ft.com/content/c615dbba-cb22-11e9-af46-b09e8bfe60c0

Klarna. (2024). *Klarna AI assistant handles two-thirds of customer service chats in its first month.* Opened on July 15, 2024, available at https://www.klarna.com/international/press/klarna-ai-assistant-handles-two-thirds-of-customer-service-chats-in-its-first-month/

Leavenworth, S. (2016, June 15). *China's 'naked loans' force female students to bare all in return for more cash.* Opened on December 4, 2023, available at https://www.theguardian.com/world/2016/jun/15/chinas-naked-loans-force-female-students-to-bare-all-in-return-for-more-cash

Liddell, H. G., & Scott, R. (1889). An intermediate Greek-English Lexicon. In H. G. Liddell & R. Scott, *An intermediate Greek-English Lexicon.* Clarendon Press. Opened on December 4, 2023, available at https://www.perseus.tufts.edu/hopper/text?doc=Perseus%3Atext%3A1999.04.0058%3Aentry%3Dlo%2Fgos

Llewellyn, D. T., Steare, R., & Trevellick, J. (2014). *Virtuous banking placing ethos and purpose at the heart of finance.* ResPublica.

Lozano, L. (2022). *ESBG/WSBI—Number of unbanked adult EU citizens more than halved in the last four years.* Opened on June 13, 2024, available at https://www.wsbi-esbg.org/number-of-unbanked-adult-eu-citizens-more-than-halved-in-the-last-four-years/

Manning, L. (2022). *Deposit multiplier: Definition, how it works, and calculation.* Opened on June 13, 2024, available at https://www.investopedia.com/terms/d/deposit_multiplier.asp

Marco Da Rin, T. H. (2002, August). Banks as catalyst for industrialisation. *Financial markets group, the London school of economics and political science, 11*(4), 4–7.

Mazzucato, M. (2019a). In M. Mazzucato, *The value of everything—Making and taking in the global economy.* Pengiun Books.

Mazzucato, M. (2019b). In *The value of everything—Making and taking in the global economy*. Pengiun Books.

McConnell, J. J., & Buser, S. A. (2011). The origins and evolution of the market for mortgage-backed securities. *Annual Review of Financial Economics, 3*, 173–192.

Merz, A., Willems, I., & van Wingerde, K. (2021). *Tussen eed en tucht—Ervaringen met bankierseed en tuchtrecht in de dagelijkse praktijk*. Opened on March 30, 2024, available at https://www.nvb.nl/media/4604/rapport-tussen-eed-en-tucht_mei-2021.pdf

Milne, R. (2018, September 26). *Financial times: Danske Bank whistleblower was British executive in Estonian branch*. Opened on March 17, 2024, available at https://www.ft.com/content/32d47fd8-c18b-11e8-8d55-54197280d3f7

Mt Hood Community College. (2019). *Pathos (emotions, values)*. Opened on May 5, 2024, available at https://mhcc.pressbooks.pub/wr122/chapter/6/

nederlandse vereniging van banken. (2015). *Bankierseed / belofte*. Opened on March 30, 2024, available at https://www.nvb.nl/themas/stabiele-en-gezonde-banken/toekomstgericht-bankieren/bankierseed-belofte/

Nickel. (2024). *Nickel.eu*. Opened on July 1, 2024, available at https://nickel.eu/en-be

Olthoorn, S. (2024). *FD—Volksbank-filosoof Joris Krijger: 'AI maakt altijd fouten. Hoeveel vinden we acceptabel?'*. Opened on July 30, 2024, available at https://fd.nl/tech-en-innovatie/1523449/volksbank-filosoof-joris-krijger-hoe-zorg-je-dat-inzet-van-ai-past-bij-jouw-waarden

PR Newswire. (2024). *Study reveals annual cost of financial crime compliance totals $61 billion in the United States and Canada*. Opened on July 9, 2024, available at https://www.prnewswire.com/news-releases/study-reveals-annual-cost-of-financial-crime-compliance-totals-61-billion-in-the-united-states-and-canada-302064989.html

Rabinowitz, H. (2022). *CNN—Danish bank pleads guilty to multi-billion dollar fraud scheme on U.S. Banks*. Opened on March 10, 2024, available at https://edition.cnn.com/2022/12/13/business/danske-bank-justice-department-fraud/index.html

Rabobank. (2024a). *Rabo SmartBuilds*. Opened on September 5, 2024, available at https://www.rabobank.nl/over-ons/smartbuilds

Rabobank. (2024b, March 7). *Our impact in 2023*. Opened on September 10, 2024, available at https://media.rabobank.com/m/65aa03e2359b413d/original/Our-Impact-in-2023.pdf

Rapp, C. (2023). *"Aristotle's Rhetoric", The Stanford encyclopedia of philosophy (Winter 2023 edition)*. Opened on July 02, 2024, available at https://plato.stanford.edu/archives/win2023/entries/aristotle-rhetoric/

Reuters. (2023a). *Estonia charges six ex-Danske Bank branch staff in money laundering case*. Opened on March 10, 2024, available at https://www.reuters.com/world/europe/estonia-charges-six-ex-danske-bank-branch-staff-money-laundering-case-2023-04-17/

Reuters. (2023b). *Greece to start sale of National Bank stake in coming weeks—Sources*. Opened on November 5, 2023, available at https://www.reuters.com/markets/deals/greece-start-sale-national-bank-stake-coming-weeks-sources-2023-11-02/

Rosemain, M. (2024). *Reuters—BNP settles Swiss franc mortgage dispute*. Opened on April 10, 2024, available at https://www.reuters.com/legal/bnp-paribas-settles-swiss-franc-mortgage-dispute-2024-01-02/

Rovera, C. (2022). Ethics in banking—Is it possible? In C. Rovera, *Ethiscs in banking—Is it possible?* (p. 7). Springer Nature Switzerland.

Saks Frankel, R. (2021, July 27). *History of credit cards*. Opened on December 13, 2023, available at https://www.forbes.com/advisor/credit-cards/history-of-credit-cards/

Sarafidis, V., Panagiotelis, A., & Theodore, P. (2017). When did it go wrong? The case of Greek Sovereign debt. In *The Greek debt crisis—In quest of growth in times of austerity* (pp. 1–31). Springer Nature.

Saravia, F. (2022, December). *Banking in Europe: EBF facts & figures 2022*. Opened on December 14, 2023, available at https://www.ebf.eu/wp-content/uploads/2022/12/Banking-in-Europe-EBF-Facts-and-Figures-2022.-FINAL.pdf

Schepis, G. (2022). *National law review: Danske Bank scandal proves need for AML whistleblower law*. Opened on March 17, 2024, available at https://www.natlawreview.com/article/danske-bank-scandal-proves-need-aml-whistleblower-law

Schoppen, E. (2017). *YouTube—The neuroscience behind brand trust—Lecture by Erik Schoppen*. Opened on March 30, 2024, available at https://www.youtube.com/watch?v=u7xqS2nNYUo&t=738s

Serrure, B. (2024, September 19). *De Tijd - ING gaat klimaatachterblijvers dumpen als klanten*. Opened on September 21, 2024, available at https://www.tijd.be/ondernemen/banken/ing-gaat-klimaatachterblijvers-dumpen-als-klanten/10565065

Skydsgaard, N., & Anthony, D. (2021). *Reuters—Danske CEO bows out over Dutch ABN money laundering inquiry*. Opened on October 30, 2024, available at https://www.reuters.com/business/danske-ceo-bows-out-over-dutch-abn-money-laundering-inquiry-2021-04-19/

Snappi Bank. (2024). *Snapp—Home—About us*. Opened on August 15, 2024, available at https://www.snappibank.com/en/about-us/

Statista.com. (2023, August). *Total value of investments into fintech companies worldwide from 2010 to H1 2023*. Opened on December 15, 2023, available at https://www.statista.com/statistics/719385/investments-into-fintech-companies-globally/

Suy, P. (2024, September 27). *De Tijd - Commerzbank-soap doet bankiers dromen van nieuwe Europese overnamegolf*. Opened on September 27, 2024, available at https://www.tijd.be/ondernemen/banken/commerzbank-soap-doet-bankiers-dromen-van-nieuwe-europese-overnamegolf/10566310.html

Taylor, M. (2024). *Will the US embrace Open Banking?* Opened on May 5, 20424, available at https://cube.global/resources/blog/will-the-us-embrace-open-banking

The Cambridge Dictionary. (2023, November 17). *Cambridge dictionary—Ethics*. Available at https://dictionary.cambridge.org/dictionary/english/ethics

Thompsett, L. (2024). *HSBC sets out bold net zero transition plan.* Opened on September 5, 2024, available at https://fintechmagazine.com/articles/hsbc-sets-out-bold-net-zero-transition-plan

Toole, P. (2014, April 8). *From shopping to space travel, how the mainframe changed our world.* Opened on December 14, 2023, available at https://www.forbes.com/sites/ibm/2014/04/08/from-shopping-to-space-travel-how-the-mainframe-changed-our-world/

Tupper, H., & Ellis, S. (2024, March 22). *4 Experiments to encourage employees' career progress.* Opened on August 20, 2024, available at https://hbr.org/2024/03/4-experiments-to-encourage-employees-career-progress

UBS. (2022). *The history of digital banking.* Opened on July 1, 2024, available at https://www.ubs.com/ch/en/wealth-management/womens-wealth/magazine/articles/history-of-digital-banking.html

UNFCCC. (2024). *The Paris Agreement—What is the Paris Agreement?* Opened on July 22, 2024, available at https://unfccc.int/process-and-meetings/the-paris-agreement

Van Der Linden, M. J. (2022). Design guidelines for the monetary and financial system in the digital age. In M. J. Van Der Linden, *Design guidelines for the monetary and financial system in the digital age* (p. 169). Delft University of Technology.

van der Vorst, R. (2023). *De Toekomst is Eindeloos—Een open blik op digitalisering.* Uitgevery Ten Have.

Vanderhoydonk, K., & van Beek, R. (2019). Regtech: The story that continues. In *#REGTECH BLACKBOOK* (p. 16). Life&Finance B.V.

Vonck, S. (2024). *Investment Officer—'Een bankier mag geen groot ego hebben'.* Opened on June 13, 2024, available at https://www.investmentofficer.be/nl/nieuws/een-bankier-mag-geen-groot-ego-hebben?utm_source=mailchimp&utm_medium=email&utm_campaign=nieuwsbrief&utm_source=Investment+Officer+Belgium+-+News&utm_campaign=bf69b4ec76-'Een+bankier+mag+geen+groot+ego+hebben'+Me

Vorster, A. (2024a). *The Banking Scene - The AI Stratego—Defining the rules for banking.* Opened on August 15, 2024, available at https://thebankingscene.com/opinions/the-ai-stratego-defining-the-rules-for-banking/

Vorster, A. (2024b, June 27). *The Banking Scene - Generative AI in Benelux banking—Opportunities, challenges and future outlook.* Opened on July 30, 2024, available at https://thebankingscene.com/resources/generative-ai-in-benelux-banking-white-paper/

Vorster, A. (2024c, June 27). *Generative AI in Benelux banking—Opportunities, challenges and future outlook.* Opened on August 30, 2024, available at https://thebankingscene.com/resources/generative-ai-in-benelux-banking-white-paper-2/

Vynck, G. D. (2024). *The Washington Post—A year after imploding, Silicon Valley Bank tries to make a comeback.* Opened on April 21, 2024, available at https://www.washingtonpost.com/technology/2024/03/14/silicon-valley-bank-collapse-acquisition-svb/

Wass, S. (2020). *S&P Global: Danske Bank receives fine for misleading customers in Flexinvest Fri scandal*. Opened on July 3, 2024, available at https://www.spglobal.com/marketintelligence/en/news-insights/latest-news-headlines/danske-bank-receives-fine-for-misleading-customers-in-flexinvest-fri-scandal-61081980

Wlodarczak-semcwuk, A. (2023). *Reuters—EU top court backs consumers in Polish FX mortgage case*. Opened on November 15, 2023, available at https://www.reuters.com/business/finance/eu-top-court-backs-consumers-polish-fx-mortgage-case-2023-06-15/

Wood, J., & Berg, P. (2011). *Rebuilding trust in banks*. Opened on June 14, 2024, available at https://news.gallup.com/businessjournal/148049/rebuilding-trust-banks.aspx

Woodman, S. (2020, September 21). *HSBC moved vast sums of dirty money after paying record laundering fine*. Opened on December 14, 2023, available at https://www.icij.org/investigations/fincen-files/hsbc-moved-vast-sums-of-dirty-money-after-paying-record-laundering-fine/

World Commission on Environment and Development. (1987, March 20). *Report of the world commission on environment and development: Our common future*. Opened on August 5, 2024, available at https://sustainabledevelopment.un.org/content/documents/5987our-common-future.pdf

Zhu, X. (2024). *Open banking and changing relationships between the Chinese banks an FinTech providers*. Opened on May 5, 2024, available at http://etheses.dur.ac.uk/15319/5/Open_banking_-_MRes_Thesis-Xiaoxi_ZHU.pdf?

Ziady, H. (2023). *CNN—Why Silicon Valley Bank collapsed and what it could mean*. Opened on April 21, 2024, available at https://edition.cnn.com/2023/03/13/investing/silicon-valley-bank-collapse-explained/index.html

Index

R. Coeckelbergs, *A New Ethos in Banking*,
https://doi.org/10.1007/978-3-031-78234-3